CIVIL WAR IN FAYETTE COUNTY WEST VIRGINIA

by Tim McKinney

WINTER QUARTERS.
Built by Col. R. B. Hayes in the Valley of the Kanawha, and occupied by himself and family in the winter of 1862-63.

FROM HOWE'S *HISTORY OF OHIO*

Charleston, West Virginia

©1988 Tim McKinney.

All rights reserved. No part of this book may be reproduced in any form
or means, electronic or mechanical, including photocopying, recording,
or by any information storage and retrieval system,
without permission in writing from the publisher.

Library of Congress Control Number: 88-90971
ISBN 13: 978-1-891852-91-6
ISBN 10: 1-891852-91-4

12 11 10 9 8 7

Printed in the United States of America

Distributed by:

West Virginia Book Co.
1125 Central Avenue
Charleston, WV 25302

www.wvbookco.com

Introduction

Fayette County was formed by an act of the General Assembly of Virginia, February 28, 1831, from Kanawha, Nicholas, Greenbrier and Logan counties. The county was named in honor of General LaFayette, the distinguished Frenchman of Revolutionary fame. The first county-seat of Fayette was located at New Haven, in Mountain Cove district. It was later moved to Vandalia, which was named for Abraham Vandal, and at whose house court was held until the completion of the public building. The name of the site was changed to Fayetteville in 1837, since which time it has been the seat of justice.

As a native of Fayette County I feel a special attachment to its residents, past and present. Having had a life-long interest in the American Civil War, it seems only natural that I should write a history of the Civil War in Fayette County.

This county was occupied by troops of the Blue or Gray continuously from April 1861 to April 1865. The overwhelming majority of men who enlisted from Fayette County joined the Southern Army. Admittedly, this work is focused more on their activities during 1861, than on those of the Federal troops, who were not native Fayette Countians, but who came here predominately from Ohio as an invading army, or as liberators, depending on your point of view.

I have long felt that West Virginia Confederates generally, have not been "given their due," and I hope this work will, at last in some small way, begin to correct that. The counties that make-up West Virginia sent at least 11,000 troops to the Confederacy, and 29,000 to the Union Army. This is a larger percentage of Confederate versus Federal troops than enlisted from the State of Kentucky, though many people consider Kentucky to be more of a "Confederate State."

In compiling this work I have attempted whenever possible to let the actual participants "speak for themselves," through the use of manuscript and published material. It is hoped that in this way I have been able to impart to some degree, the depth of struggle and range of emotions experienced by thousands of men, women, and children, during the Civil War.

Much of the information presented here has never been published before and will, I think, be quite a surprise to many history buffs. It has been a pleasure for me to assemble this material and I welcome correspondence with anyone who has, or needs, information concerning the Civil War in West Virginia.

During the many months this work was underway I was very encouraged to note the many cheerful and friendly expressions of sincere interest in the progress and success of my undertaking. Though the assistance of many people is acknowledged in another section, several people merit special mention.

Dr. Otis K. Rice of the History Department at West Virginia Institute of Technology, Montgomery, W. Va. Dr. Rice volunteered to edit my manuscript, loaned me various items from his Civil War files, and offered advice along the way. He is a good friend who has always had time for me and has never made me feel like the novice that I am.

Terry Lowry of South Charleston, W. Va., whose Civil War books helped feed the fires of curiosity. Terry and several other friends encouraged me over the last few years to write a Civil War book. Without Terry's friendship and multifarious support this work would have been less complete.

Bob and Christine Beckelheimer of Oak Hill, W. Va., who opened their home and Civil War files to me. Their collection of the Official Records were a convenient and ready source of information to which they allowed me unlimited access.

Jody Mays of Beaver, W. Va., who loaned me numerous items from his files and offered support throughout this work. As a co-founder of the "Trans-Allegheny Historical Association," Jody and his friends have done a fine job preserving the history of southern West Virginia through their historic surveys and photography.

Ed and Nina Clark of Kanawha Falls, W. Va., who are certainly the best friends a man could have. They have offered me their consistent friendship and assistance in all my pursuits of the last few years, for which I am eternally grateful.

My parents, Jerry and Barbara McKinney of Indianapolis, Indiana. Without their love and support much of my life would seem empty.

My brothers, Jerry and Mark, and my sister Jane. Also my grandparents, Kermit and Eula Taylor of Robson, W. Va., who have shown me their love and support over all these years.

My uncle, Charlie Taylor of Robson, W. Va., who has been my most active relic hunting partner of the last few years, and who has patiently searched many West Virginia Civil War sites with me. Sometimes we got "famous," and sometimes we flopped, but we have never stopped trying.

Last but never least, I want to thank God for bringing me this far.

Acknowledgments

One of the most enjoyable aspects of doing the research for this work was the many friends I made in my travels to various libraries and archives, and in correspondence to areas where time would not allow me to travel. I now have new friends and acquaintances in areas from Pennsylvania to Louisiana, Virginia to Ohio.

In compiling this list of people to whom I am much obliged, I have attempted to include everyone and I sincerely apologize for anyone I may have accidentally omitted.

Linda Allen—Vining Library, West Virginia Tech, Montgomery, W.Va. Ann Alley—Tennessee State Library & Archives, Nashville, Tenn. Ofelia Allexander—Vining Library, West Virginia Tech, Montgomery, W. Va. Linda Bailey—Cincinnati Historical Society, Cincinnati, Ohio. Bob & Chris Beckelheimer—Fayette Co. Historical Society, Oak Hill, W. Va. Dr. Arthur W. Bergeron Jr.—Louisiana Department of Culture & Tourism, Baton Rouge, Louisiana. Leona Gwinn Brown—Arbovale, W. Va. Nina & Ed Clark—Kanawha Falls, W. Va. Stan Cohen—Pictorial Histories Publishing Co., Charleston, W. Va. Dr. James A. Davis—Shenandoah College & Conservatory, Winchester, Va. Lacy W. Dick—The Valentine Museum, Richmond, Va. Jack & Kay Dickinson—Barboursville, W. Va. John & Bonnie Estes—Rainelle, W. Va. George Fisher—Boomer, W. Va. Eva Fitzwater & family—Lookout, W. Va. Patricia Gantt—Wilson Library, University of North Carolina, Chapel Hill, N.C. R. P. Gravely—Martinsville, Va. A. N. Gwinn—Grand Rapids, Michigan. E. M. Hennen Jr.—Mississippi Dept. of Archives & History, Jackson, Mississippi.

Richard Holloway—Louisiana State Dept. of Archives, Baton Rouge, La. Janice Kessler—Charleston, W. Va. William C. Ledbetter—Murfreesboro, Tenn. Bonnie J. Linck—Ohio Historical Society, Columbus, Ohio. Terry Lowry—South Charleston, W. Va. Jody Mays—Beaver, W. Va. Dennis & Ann McCutcheon—Lookout, W. Va. David Miles—Charmco, W. Va. Don Mindemann—Dover, Pa. Ron Mink—West Virginia Tech, Montgomery, W. Va. Chaddra A. Moore—Tennessee State Archives, Nashville, Tenn. Bob Mount—WV. Archives, Charleston, WV. Aubrey Musick—Gauley Bridge, W. Va. Staff—Ohio Historical Society, Columbus, Ohio. Don Pomeroy—Smithers, W. Va. Charlotte Ray—Georgia Dept. of Archives & History, Atlanta, Ga. Dr. Otis K. Rice—History Dept., W. Va. Tech, Montgomery, W. Va. Dr. James I. Robertson Jr.—Virginia Polytechnic Institute, Blacksburg, Va. Wilma Sargent— Boomer, W. Va. Philip Shiman—Duke University Library, Durham, N. C. Mike Smith & family—Droop Mtn. State Park, Hillsboro,

W. Va. Thomas A. Smith—Rutherford B. Hayes Library, Freemont, Ohio. Richard Spearen—Fayetteville, W. Va. Jim Spounagle—Robson, W. Va. Bob Sprayberry—Fayette Co. Georgia Historical Society, Fayetteville, Ga. Charles Taylor—Robson, W. Va. Roger Thompson—130 W. 11th Ave., Huntington, W. Va. Wayne Varner & family—Beverly, W.Va. James & Virginia Vaughan—Martin, Tenn. Staff—Virginia Historical Society, Richmond, Va. Staff—Virginia State Library—Richmond, Va. Staff—West Virginia State Archives, Charleston, W. Va. Michael J. Winey—U. S. Army Military History Institute, Carlisle Barracks, Pa. Barbara Workman—Vining Library, West Virginia Tech, Montgomery, W. Va. Geraldine Workman—Lansing, W. Va.

I regret that several of my friends and relatives did not live to see this work. My grandmother, Tilda McKinney of Jackson, Ohio. My cousin, Pearl Spounagle of Robson, W. Va., and my great uncle, William "Benny" Taylor of Robson, W. Va.

■ Dedication

Dedicated to the men in Blue & Grey who enlisted from Fayette County. Among them were my great great grandfathers, Pvt. Joseph V. Rollins, 22nd Virginia Infantry C.S.A., and Cpt. William Taylor, organizer of the local Union militia and commander of the Gauley Bridge ferryboat in 1863.

Contents

Introduction 3
Acknowledgments 5
CHAPTER ONE: Early Military Activity 11
CHAPTER TWO: The Destruction of Gauley Bridge 27
CHAPTER THREE: The Scene of Action 37
CHAPTER FOUR: Robert E. Lee in Fayette County 65
CHAPTER FIVE: The Siege of Gauley Bridge 101
CHAPTER SIX: The Long Arm of Lincoln 115
CHAPTER SEVEN: Civilian Prisoners and Refugees 133
CHAPTER EIGHT: Battle of Fayetteville 147
CHAPTER NINE: Winter Quarters 171
CHAPTER TEN: From Fayetteville to Appomattox 183
APPENDIX: The Dixie Rifles 193
Notes 204
Bibliography 215
Index 221

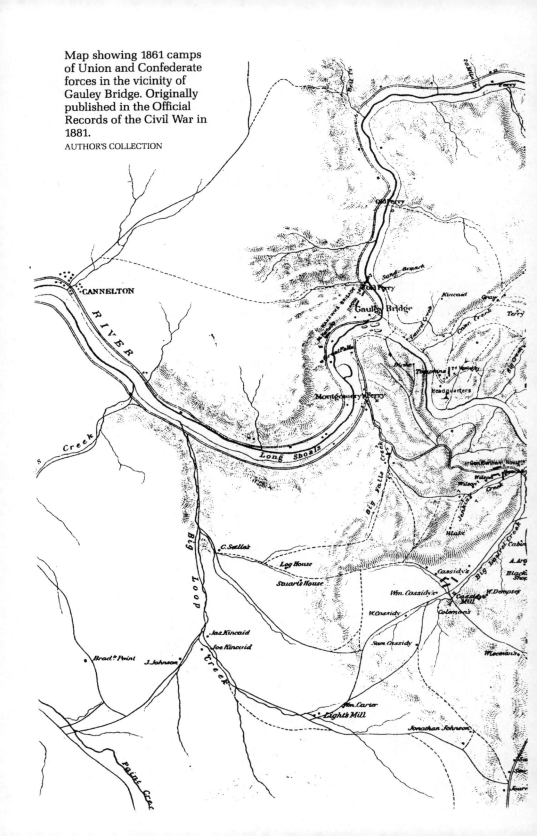

Map showing 1861 camps of Union and Confederate forces in the vicinity of Gauley Bridge. Originally published in the Official Records of the Civil War in 1881.
AUTHOR'S COLLECTION

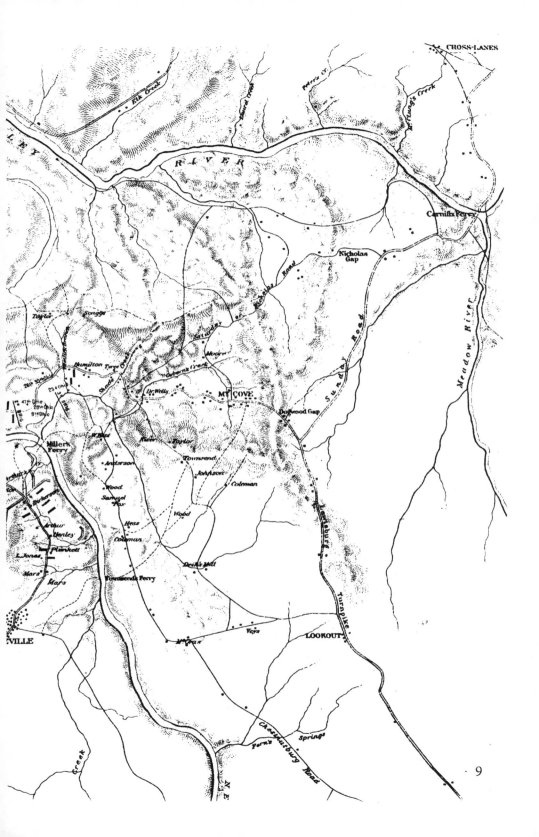

Colonel John McCausland—36th Virginia Infantry C.S.A. He was a dedicated Southern soldier and remained an "unreconstructed rebel" until his death in 1927. COURTESY SWV

■ *Chapter One*

Early Military Activity

Jim Comstock, publisher of the West Virginia Hillbilly, once wrote that when the Civil War began, April 12, 1861, the United States was split into three parts; the North, the South, and West Virginia. Mr. Comstock's assessment of the situation is correct. West Virginia falls into the category of so-called border States. Situated somewhere between the northern and southern states, the residents of Trans-Allegheny Virginia were strongly divided in their loyalties. Northern and western counties of West Virginia were largely for the Union, while southern and eastern counties were almost entirely in favor of a Southern confederacy.

The majority of people in Fayette County were in sympathy with the Southern cause. Slavery was not really a factor in their decision. The 1850 census of Fayette County listed a total white population of 3,782, 17 free colored citizens, and 156 slaves, who made up just 4% of the total population. To Confederate sympathizers in Fayette County and many other areas of the South, the question was one of states rights. They felt that their homes and property were being threatened by northern fanatics. Henry L. Gillispie, Fayette and Raleigh County delegate in the Virginia convention of 1861, voted for its secession ordinance. Fayette County sent no delegates to the First Wheeling Convention on May 13, 1861, or to the second convention held there on June 11. The residents of Fayette County wanted no part in the proposed new state of West Virginia.[1]

Very few people thought the war would last more than a few weeks, and no one could foresee the major role Fayette County would play in the war effort of Trans-Allegheny Virginia. Gauley Bridge, as the head of navigation in Fayette County and surrounding territory, was of decided military importance. Here the New and Gauley Rivers merge to form the great Kanawha. The James River and Kanawha Turnpike paralleled New River to the east. Another road extended from Gauley to Summersville, a distance of about thirty miles, with a side road to Cross Lanes and Carnifax Ferry. From Summersville, this road continued northward to Sutton and Weston, thus making a line of communications between northwest Virginia and the Kanawha Valley. The Giles, Fayette, and Kanawha Turnpike, extended from Kanawha Falls through Fayetteville to Flat Top Mountain, another strong position, thence to the Narrows of New River, and on to the important Virginia and Tennessee Railroad.

A meeting of Confederate sympathizers held at Gauley Bridge, April 27, 1861, unanimously resolved:

> 1st. That an immediate response to the proclamation of the Governor of Virginia should be made by the organization of a volunteer company of riflemen, pledged to defend the honor and interest of the State, and with this view to cooperate with other military companies along the route of the Kanawha Valley.
>
> 2ndly. That until arms and equipment of suitable character can be obtained, each citizen shall provide his own rifle and equipments, and be mustered at such time and place as may be appointed with the least possible delay.
>
> 3rdly. That the Colonel of the 142 regiment be requested to superintend the election of officers above notified on the subscription of the requisite number of names.[2]

On April 29, Robert E. Lee commissioned Kanawha Valley resident John McCausland, a Lieutenant Colonel of volunteers, and placed him in charge of Confederate recruitment efforts in the Kanawha Valley.[3] McCausland was a logical choice for the assignment. He had been raised in the Kanawha Valley and had received his early education at the Buffalo Academy. He had a military background that included graduation from the Virginia Military Institute, Lexington, Virginia. After his graduation he served as assistant professor of mathematics at V.M.I.

Colonel Christopher Q. Tompkins—22nd Virginia Infantry. His home at "Gauley Mount," two miles east of Gauley Bridge was taken over as a headquarters and main camp for the Union Army.
COURTESY VIRGINIA STATE LIBRARY

On May 3, Robert S. Garnett, Adjutant General of Virignia forces, commissioned Christopher Q. Tompkins, another local resident, Colonel of Virginia volunteer forces in the Kanawha Valley.[4] Tompkins was told that McCausland would assist him. This was another logical choice as Tompkins was a highly experienced military man, and a respected resident of the Kanawha Valley. His home was three miles east of Gauley Bridge, and was known as Gauley Mount. Today the site is the location of the Hawks Nest Golf Course. Gauley Mount was a "varitable showplace, with slaves, a library, a seventy foot long barn, and a vineyard of 800 vines." Mrs. Tompkins was a "cultured Richmond belle," and liked to make frequent trips to Richmond. Before the war Colonel Tompkins had been mining agent and superintendent of coal mines in Fayette and Kanawha Counties. He was forty-seven years old and probably would not have accepted his assignment had he not been under the impression that the war would be of short duration.[5]

A special meeting of the county court was held at Fayetteville on May 13, and an appropriation of $5,000 was made "For the purpose of uniforming and equipping the volunteer forces of this county, also for the support of destitute families of those who have or may volunteer their services in defense of the state." A special police force was appointed for the different magisterial districts. The duty of this force was to watch over the several districts and arrest all persons who were believed to be engaged in inciting insurrection or rebellion against the state. Appointed captains of the special police forces were: District No. 1, James Muncy; District No. 2, E. B. Bailey; District No. 3, Samuel Lewis; and District No. 4, Frances Tyree.[6] On Saturday, May 18, a general muster of Confederates was held at Fayetteville. Brigadier General Alfred Beckley was present, and during the three days' training of officers, was very effective imparting military instruction. Beuhring H. Jones, who was soon to organize the "Dixie Rifles," also addressed the soldiers and citizens. At the conclusion a call was made for volunteers and about fifty men stepped forward. The "Kanawha Rangers," under Captain Charles I. Lewis, were also present. The courthouse, public buildings, and homes were decorated, and every one partook of a dinner at the courthouse.[7]

Several Confederate camps of instruction were established in the Kanawha Valley. The largest of these was Camp Tompkins, which was located at present day St. Albans, and was situated west of the mouth of Coal River, near Tacketts Creek and the community of Coalsmouth. This military camp became the hub of Virginia volunteer forces, as recruits poured in on a near daily basis.[8] Though Confederate recruitment efforts in the Kanawha Valley had not been as successful as anticipated, about 850 men assembled at Camp Tompkins by June 4.[9] Many of the new recruits had responded to an advertisement which appeared in the *Kanawha Valley Star*. It read:

Brigadier General Alfred Beckley. He organized the 27th Brigade Virginia Militia. COURTESY MRS. M.M. RALSTEN, BECKLEY, WV

April 26, 1861: The mountains of Trans-Allegheny are filled with able bodied men, men accustomed from their youth to bear arms, every one of whom has one or more rifles in his cabin, and all of whom are first rate marksmen. These men number legions, and a little drilling would make them the best of soldiers. Should the abolishionist of Ohio send an invading army into Western Virginia, not a soldier among them will ever return alive. The mountain boys will shoot them down as dogs.

Still others responded to a proclamation issued by Colonel Tompkins:

Charleston, Kanawha County, Va. May 30, 1861 Men of Virginia, Men of Kanawha, to Arms

The enemy has invaded your soil and threatens to overrun your country under the pretext of protection. You cannot serve two masters. You have not the right to repudiate allegiance to your own State. Be not reduced by his sophistry or intimidated by his threats. Rise and strike for your firesides and altars. Repel the aggressors and preserve your honor and rights. Rally in every neighborhood with or without arms. Organize and unite with the sons of the soil to defend it. Report yourselves without delay to those nearest you in military postion. Come to the aid of your father, brothers, and comrades in arms at this place, who are here for the protection of your mothers, wives, and sisters. Let every man who would uphold his rights turn out with such arms as he may get and drive the invader back.

C. Q. Tompkins,
Colonel, Virginia Volunteers, Commanding[10]

Living in the community of Coalsmouth near Camp Tompkins was a young woman, Victoria Hansford Teays. Victoria experienced the war first hand and left an excellent diary of the events she witnessed from 1861 to 1864 on the "home front" in the Kanawha Valley. She wrote:

The Spring of 1861 came as all other Springs with its sunshine, its birds, and its flowers, yet we hailed it not with joyousness for far away could be heard the mutterings of the storm which was soon to sweep over our beautiful valley, with its blighting power, crushing out the joy from thousands of hearts, making desolate thousands of homes. Oh, who that has only heard of an armed foe invading a country can realize for an instant the horror and privations which they bring.

Six of our young men went to Charleston to volunteer their services to the state: C. M. Hansford, S. T. Teays, N. B. Brooks, Charlie Turner, Theodore Turner, and Thomas Grant. Only six, but they were strong, healthy, patriotic and brave. The citizens of Coalsmouth gathered around the wagon to tell them goodbye, and wish them "godspeed," and tears were not only shed by their mothers and sisters, but many a true, brave, son of Virginia, wept over the offering we were about to make.[11]

Confederate activity in the Kanawha Valley had not escaped notice by the Union Government. General George B. McClellan issued a proclamation to the people of Western Virginia on May 26:

> Headquarters Department of the Ohio,
> Cincinnati, May 26, 1861
>
> To the Union Men of Western Virginia:
>
> Virginians: The General Government has long enough endured the machinations of a few factious rebels in your midst. Armed traitors have in vain endeavored to deter you from expressing your loyalty at the polls. Having failed in this infamous attempt to deprive you of the exercise of your dearest rights, they now seek to inaugurate a reign of terror, and thus force you to yield to their schemes, and submit to the yoke of the traitorous conspiracy dignified by the name of Southern Confederacy.
>
> They are destroying the property of citizens of your State and ruining your magnificent railways. The General Government has heretofore carefully abstained from sending troops across the Ohio, or even from posting them along its banks, although frequently urged by many of your prominent citizens to do so. I determined to await the result of the late election, desirous that no one might be able to say that the slightest effort had been made from this side to influence the free expression of your opinion, althought the many agencies brought to bear upon you by the rebels were well known.
>
> You have now shown, under the most adverse circumstances, that the great mass of the people of Western Virginia are true and loyal to that beneficent Government under which we and our fathers have lived so long. As long as the result of the election was known the traitors commenced their work of destruction. The General Government cannot close its ears to the demand you have made for assistance. I have ordered troops to cross the river. They come as your friends and brothers, as enemies only to the armed rebels who are praying upon you. Your homes, your families, and your property are safe under our protection. All your rights shall be religiously respected.
>
> Notwithstanding all that has been said by the traitors to induce you to believe that our advent among you will be signalized by interference with your slaves, understand one thing clearly—not only will we abstain from all such interference, but we will, on the contrary, with an iron hand, crush any attempt at insurrection on their part. Now that we are in your midst, I call upon you to fly to arms and support the General Government.
>
> Sever the connection that binds you to the traitors. Proclaim to the world that the faith and loyalty so long boasted by the Old Dominion are still preserved in Western Virginia, and that you remain true to the stars and stripes.
>
> Geo. B. McClellan,
Major-General, Commanding[12]

On June 3, the Fayetteville Rifles left Fayetteville on their way to Charleston. When they reached Charleston a few days later they were quartered in a Yankee sawmill. Then arose the desperate problem of

William F. Bahlmann of the Fayette Rifles—22nd Virginia Infantry.
COURTESY LARRY LEGGE, BARBOURSVILLE, WV

arming the men with proper weapons, as described by William F. Bahlmann of the 22nd Virginia Infantry:

> After a day or two, we marched to a store in the business part of town to receive our guns. We had been promised rifles, had volunteered as riflemen and had Fayetteville Rifles inscribed on our flag. Imagine our astonishment when they started to hand us smoothbore muskets. Being the Orderly Sergeant, I was the first man to receive a gun. I was unwilling to start a mutiny and perhaps breakup a fine company, but your Uncle Joel, the second sergeant, refused to take a gun and then the trouble began. We want rifles. Look at our flag. Rifles. Rifles. I never saw men so thoroughly exasperated. Some of the people told me that they were frightened. The men made a speech and a compromise was affected . . . [13]

Troops organized in Fayette County and sent into the Kanawha Valley included:
The Dixie Rifles, Captain B. H. Jones
The Fayetteville Rifles, Captain Robert A. Bailey
The Mountain Cove Guard
The Fayette Rangers, Captain William Tyree
Other Confederate troops organized in and near the Kanawha Valley included:
The Kanawha Riflemen, Captain George S. Patton, organized in 1856.

The Charleston Sharpshooters, Captain John S. Swann, organized 1859.
The Border Guards, Captain Albert J. Beckett
The Buffalo Guards, Captain William E. Fife, organized 1859
The Coal River Rifle Company, Captain Thomas A. Lewis, organized 1859
The Border Rangers, Captain Albert Gallatin Jenkins
The Kanawha Artillery, Captain John Peter Hale
The Elk River Tigers, Captain Thomas B. Swann
The Logan Riflemen, Captain Charles J. Stone
The Logan Wildcats, Captain Henry M. Beckley
The Chapmanville Daredevils
The Boone County Rangers, Captain James W. McSherry
The Sandy Rangers, Wayne County Cavalry, Captain James M. Corns
The Putnam County Border Riflemen, Captain Andrew Russell Barbee
The Nicholas Blues, Captain Winston Shelton[14]

By the early days of June 1861, the Confederate Government was having doubts that state volunteer forces would prove an effective fighting force. It wanted to be sure the soldiers were properly trained and led by a man who could be popular with both the troops and the inhabitants of the area. Accordingly, on June 6, Adjutant and Inspector General of Virginia, Samuel Cooper, appointed ex-Governor of Virginia, Henry Alexander Wise, Brigadier General of Provisional forces. Wise was ordered to proceed from Richmond to the Kanawha Valley, collecting what forces he could along the way.[15]

Henry A. Wise was born at Drummondtown, Virginia, December 3, 1806. He was a graduate of Washington College, Pennsylvania, studied law and was admitted to the bar in 1828. He served as Democratic governor of Virginia from 1856 to 1860, during John Brown's raid on Harpers Ferry. The final and most prominent act of his administration was the execution of Brown. Although he had no military training at all, he did have "remarkably correct apprehension of topography, and was quick to see the strategic value of positions."[16]

General Wise arrived at Lewisburg June 14, and wrote to Colonel Tompkins from there on June 17:

> They will, in all say 298 men, start tomorrow morning, and be at Gauley Bridge thursday evening next. There they will post themselves at Gauley Bridge and the falls, sending a detachment up to Twenty Mile Creek, and guarding the Gauley road, and the road leading from Twenty Mile Creek to the Kanawha Valley at Hughes Creek about fifteen miles below the bridge ... You will cover and guard the road from Parkersburg to Charleston at the gorges of the Pocatalico with as strong forces as you

General Henry A. Wise, former governor of Virginia. He organized the Wise Legion in 1861. His defiant attitude eventually cost him his command.
COURTESY SWV

can distribute to the points on that creek, scouring the countryside as far as your force will allow, and siezing, and making prisoners all resident enemies, with their arms & etc as above ordered. . . . If taken in any considerable numbers, they must be moved back up the valley, even as high as Fayetteville if necessary, or to any other sound locality where they can be guarded.[17]

General Wise left Lewisburg June 20, reached Gauley Bridge June 22, and wrote again to Tompkins the next day:

Gauley Bridge
June 23rd, 1861

To Col. C. Q. Tompkins
Commdg.

I arrived here at six p.m. yesterday. On the way, yesterday, Mr. Kirkpatrick handed me two dispatches from you of the 17th & 19th and later I met Mr. Williams who handed me yours to Cpt. Wise accompanied by yours to me all of the 21st insts. I find here today 275 men mustered in my Legion, and a Fayette company not mustered into service, in all about 340 men, besides a few unorganized men who are with Cpt. Buckholtz's artillery. By the evening Cpt. Brooks cavalry of about 90 men will be here and tomorrow mg the whole force will be about 430 men. I will leave two untrained companies here, say 130, with Cpt. Buckholtz to be drilled in infantry and artillery practice and to guard the Gauley Bridge, and will take on with me tomorrow two of the infantry and two of the cavalry companies of my Legion to Charleston, say 300 efficient men, well armed. When I see you I will be happy to interchange views upon all the topics of this command. I am now hurried by the messenger, Mr. Tompkins, who will take this. Saw your sons yesterday who report all well at Gauley Mtn., a kind note from your good lady which I shall be glad to avail myself of as I can, to make your home my home in this neighborhood. I ought to add that I am expecting some six to ten companies here from the east by the last of the week.

Very Respectfully
Henry A. Wise
Brig. Genl.[18]

Defensive preparations were quickly enacted at Gauley Bridge. Seven posts were established, all under the command of Major Bradfute Warwick, and most of them manned by the Dixie Rifles of Captain Beuhring H. Jones, composed of men from Fayette and adjacent counties. One post was located up New River, two up Gauley River, a fourth up Scrabble Creek, and another near the top of the precipitous mountain overlooking Gauley Bridge, the sixth near the falls, and the final one on the small island in the river, opposite the guard house and artillery camp.[19]

Confederate patriotism was running high in Fayette County, and at the June term of the county court, consisting of the justices of the peace from the various magisterial districts, the following resolutions were unanimously adopted and included in the minutes of the court:

> Whereas, our state has been invaded by a hostile army of northern fanatics and we feel bound to resist said invasion to the last extremity resolved therefore,
>
> First: That we feel it to be our duty in accordance with an act of the legislature passed January 19th, 1861, to levy on the people of the county from time to time as may be necessary to enable us to resist said invasion successfully such amount of money as we shall think practicable and expedient.
>
> Second: That we will then, after money and property are exhausted, feel it to be our duty to levy for said purpose on the credit of the county and when that also is gone, we will eat roots, and drink water and still fight for our liberty unto death.
>
> Third: That should any members of this court feel friendly to the North, we invite them or him to peacefully and civily resign their or his commission.[20]

In an attempt to speed supplies coming into his legion General Wise ordered Colonel St. George Croghan to proceed to North Carolina and procure a large quantity of muskets. Wise wrote:

> Headquarters, Wise Legion
> Gauley Bridge, June 24th.
>
> Special orders No. 10
>
> Col. Croghan having information that he can get 500 rifled muskets from Fayetteville, North Carolina, will proceed immediately to that place and forward them to me at Charleston if they can be got and in passing Lewisburg he will obtain from the Quartermaster the necessary certificate of transportation who is required to provide means therefor.
>
> By Command of
> Brig. Genl. H. A. Wise
> E. J. Harvie
> Cpt. Asst. Adjut. Genl.[21]

Though many of the soldiers in the Wise Legion had enlisted several weeks previously, they still had not been issued weapons or basic supplies such as tents. General Wise remained at Gauley Bridge until June 25, when he left for Charleston. A number of units accompanied him to Charleston, including the Richmond Light Infantry Blues, commanded by his son, Captain Obidiah Jennings Wise; The Greenbrier Riflemen; The Pig River Invincibles from Pittsylvania County; Jackson's Invincibles from the old town of Alexandria, Virginia; Cavalry of Captain John P. Brock, known as either the Valley Rangers or the Rockingham Cavalry; and also the personal staff of General Wise, including Colonel Charles Frederick Henningson, later to command Wise's second regiment.[22]

As the Wise Legion left Gauley Bridge for Charleston, a dispatch was sent to the Adjutant General of Virginia, by Brigadier General Alfred Beckley, for whom the city of Beckley is named. General Beckley had been recruiting and training militia companies from Fayette and Raleigh counties for several weeks.

> I have the honor to report, through yourself to his excellency the Governor and commander in chief, . . . I have been most successful in getting up a patriotic union of men, hitherto of various shades of opinion, for the defense of Virginia, and also in promoting the formation of volunteer companies, some now in the field, and several to march in a few days.
> Nicholas County, 129th regiment, two companies, one in camp; Fayette County, 142nd regiment, three companies, in the field; Raleigh County 184th regiment, two companies, one already gone, and one to leave Monday; Wyoming County, 190th regiment, one company; Logan County, 129th regiment, two companies; Boone County, 187th regiment, two companies, another forming; one marches today, one in camp; in all twelve companies.
> Thus the Governor will percieve that one Brigadier of Virginia Militia has attempted to fulfill his responsibilities, and is ready, whenever called upon, to take the field at the head of his brigade. . . .[23]

General Beckley was also having trouble arming and supplying the 300 troops he had by this time assembled. Some of his men were without shoes and others were in need of canteens and tents. Weapons had been issued to most of the troops but they were of the old smoothbore, flintlock variety, and not effective at distances over 100 yards. In fact, many of the muskets issued to Beckley's and Wise's troops were relics from the War of 1812. The militiamen asked for rifled muskets but they never received them, as none were available.

General Wise arrived at Charleston on Wednesday, June 26. A camp was established near the courthouse, with headquarters at the Kanawha House, a hotel operated by John Wright, at the corner of Front and Summers Streets. Several days later Wise moved his troops to the site of Camp Two Mile, located on the westside farm of Adam Brown Dickinson Littlepage, and his wife, Rebecca.

Camp Two Mile was named after the small stream which entered the Kanawha River after flowing through the Littlepage property. This camp extended from the Kanawha to the junction of the Ripley-Ravenswood road with the Point Pleasant road (present Washington Street West). Many of the soldiers pitched their tents in the apple orchards surrounding Kanawha Two Mile, which give their name to the Modern Orchard Manor housing development.[24]

General Wise found the military situation at Charleston to be in a chaotic condition. Many of the company commanders were incompetent, with the exception of a few stand-outs, such as Captain George S. Patton, Captain Albert Gallatin Jenkins, Lieutenant Colonel John

McCausland, and Colonel Tompkins. Soldiers of the Wise Legion were poorly trained and disorganized and it would soon become apparent that Wise himself was not going to be the saviour of Confederate efforts in the Kanawha Valley.

On July 2 General McClellan ordered Brigadier General Jacob Dolson Cox to proceed to Gallipolis, Ohio, and prepare an invasion of the Kanawha Valley.[25] General Cox was given command of approximately 3000 men, consisting of the First and Second Kentucky Infantry, the 12th and 21st Ohio Infantry, a troop of cavalry, and a battery of artillery.[26] McClellan was anxious to occupy the Kanawha Valley before Confederate strength there became a serious problem. The Federal Government had reliable information concerning Confederate efforts in the region and General Cox was ordered to act promptly. Preparations for the invasion were quickly made and in a few days the Wise Legion would find itself in a bad situation indeed.

Brigadier General Jacob D. Cox. He was a highly efficient and dedicated military man. His troops were camped in Fayette County many times during the war.
COURTESY MUDD LIBRARY, OBERLIN COLLEGE

On July 8, E. J. Harvie, Captain and Assistant Adjutant General for the Wise Legion, submitted a discouraging report of the Legion's status at that time.

CAVALRY

The Valley Rangers comd. by Capt. Brock, has, rank and file	49
Caskie Mounted Rangers comd. by Capt. Caskie, has rank and file	54
Total now here and in service	133
Company comd. by Capt. Tate (not arrived) is supposed to have rank and file	64
Total Cavalry reported	197

INFANTRY

Richmond Blues, comd. by Capt. Wise, has, rank and file	94
White Sulphur Rifles, comd. by Capt. Morris, has rank and file	57
Biernes' Sharpshooters, comd. by Lieut. Rowan, has, rank and file	65
Company comd. by Capt. McComas, has, rank and file	94
Company comd. by Capt. Thrasher, has, rank and file	68
Company comd. by Capt. Buster, has, rank and file	72
Jackson Avengers comd. by Capt. Jones, has, rank and file	89
Company comd. by Capt. Lowrie, has, rank and file	94
Total of Infantry in service	633
Company of Capt. Hammond, has, rank and file	64
Company comd. by Capt. Crank, has, rank and file	59
Company from Pittsylvania has rank and file	108
1st Appomattox Co. (supposed to have)	64
2nd Appomattox Co. (supposed to have)	64
Total of Infantry as reported	992
Kirby's Co. of Artillery supposed to have rank and file	50
Total of Cavalry as reported	197
Total of Infantry	992
Total of Artillery	50
Total of Brigade yet reported	1239
Total of Cavalry arrived	133
Total of Infantry arrived	633
Total of forces arrived and on duty	766
Total of Brigade arrived and on duty	766
Volunteer forces of Virginia	2097
Total under your command	2863 men

The total number of troops under Wise's command was listed at 2,863. The report went on to say that the men were entirely destitute of all the essentials necessary to make soldiers effective combatants—weapons, uniforms, tents, knapsacks, blankets and so forth, and led by incompetent officers, resulting in a near total lack of discipline and training. Harvie concluded that "In their present condition, untrained and awkward as many of them are in the use of arms, but little is to be expected of them, and whenever brought into collision with a disciplined force, it would be unwise to rely on them for attack or defense." He went on to say, however, they furnished the material for an army of hardy and capable soldiers, but much labor must be expended on them to make them such.[27]

The Confederate forces did continue to grow and soon reached an aggregate of 3,500 men. On July 12 Colonel St. George Croghan returned to Gauley Bridge from his search for supplies in North Carolina and Richmond. Upon his return he submitted a detailed report to General Wise, who had returned to Gauley Bridge from Charleston. It read:

> Gauley Bridge, Va. July 12th 1861
> Brig. Genl. Henry A. Wise:
>
> Sir
>
> In obedience to your order of June 24th, I proceeded to Fayetteville, North Carolina, to obtain arms. I succeeded in obtaining five hundred stand, together with a quantity of military equipments. I regret that my orders did not give me a carte blance to act as I thought proper in making my requisitions; as I think I could have brought a larger supply of useful articles for your service; if not a few more arms. Being met with your order for five hundred arms and ammunition; whenever I made a request for anything else, I was forced to make use of a little outside pressure, and succeeded to a certain extent, but not as far as I could wish. I obtained—
>
> | | |
> |---|---|
> | 180 percussion rifles | 320 percussion muskets |
> | 180 screw drivers | 180 screw drivers |
> | 180 wipers | 180 wipers |
> | 180 cones | 180 cones |
> | 18 ball screws | 32 ball screws |
> | 18 spring vises | 32 spring vises |
> | 18 bullet molds | |
>
> The articles not embraced in your order are—
> 6 artillery gunners haversacks
> 711 cap pouches
> 7 port vise case
> 597 cartridge box & other belts
> 385 cartridge boxes or pouches (old style)
>
> Having obtained this property I proceeded to Richmond and having seen from the nature of the country about Gauley Bridge how serviceably hand grenades might be used from the rocks overlooking the road, I made an application to Col. Gorgas of the Ordinance Department, for them; but learned they could not be obtained. I applied for permission to visit the laboratory and try a substitute. After some little time I succeeded in obtaining 100 six pound shells loaded with powder and bullets, and with wooded fuse attached, graduated to a scale of 6 seconds of time. These will produce probably greater effect than the ordinary hand grenade and are but

little heavier. At Richmond I obtained—
100 6 lb shell
3000 rifle cartridges, "minnies"
25 bell tents
25 common tents
500 soldiers caps
500 blankets

The quartermaster could not furnish shoes but promised to furnish 1000 pair in two or three days, together with 200 tents, 500 additional blankets, and other articles for which requisition was made. That the Quartermaster will have them ready I am certain of, but that they will be attended to and forwarded from Richmond, is for some time in my mind a matter of doubt, owing to a want of efficiency or mismanagement in the direction of public affairs. While at Richmond, learning that cartridge boxes were being manufactured at Staunton; I obtained an order from Col. Myers, QM General from Major Harmon, for 1000 of them. They are to be sent to this place forthwith. I saw Major Harmon and he assured me they should be forwarded within one week. They are probably on the way at this time. On arrival at Lewisburg, I obtained 13000 musket cartridges, handed over to me by Cpt. Thomas upon your requisition I believe. All the articles herein numerated I have brought to this point, and have handed them over to the proper officers. . . . I have the pleasure of informing you that I have brought with me a gallant officer, who is desireous of joining your command. He is Col. Frank Anderson, the right hand man of the late Genl. Wm. Walker of Nicaragua. He is an experienced officer and has proven himself in many desperate battles. I sincerely trust that you will find immediate use for him.

I am Sir respectfully Yours,
St. George Croghan[28]

In this chapter we have examined early military activity as it affected Fayette County for the period April 12 to July 12, 1861. It is impossible for modern man to fully comprehend the cataclysmic effect this 90-day period had on Fayette County, and the country as a whole.

Because sentiments were so strongly divided in this area, many of the local inhabitants were imprisoned by Union and Confederate forces. Anyone suspected of being in sympathy with the enemy was subject to arrest. The quiet, peaceful life that local residents were accustomed to was rapidly taken away. It didn't take long for people to realize that this was not going to be a "quick war."

■ *Chapter Two*

The Destruction of Gauley Bridge

On the day that Colonel Croghan delivered the desperately needed supplies to Gauley Bridge, the Union invasion force under General Cox was making its way up the Kanawha Valley from Point Pleasant. The Federal troops were moving up the valley in a three-pronged enveloping movement. The Second Kentucky Infantry entered by way of the mouth of Guyandotte River; the First Kentucky Infantry moved from Ravenswood to strike Wise's base at Kanawha Two Mile; the main army moved up the Kanawha River on four steamboats—*The Economy, Mary Cook, Matamora,* and *Silver Lake*. Their wagon train, which was over one mile in length, moved up the valley road under a heavy guard.[1]

The advancing columns met only light Confederate resistance and proceeded up the valley. Several small skirmishes developed along the way, culminating in a brief battle at Scary Creek, Putnam County, on July 17. After the fighting both sides withdrew from the field. Confederate commanders realized the situation and returned to claim a victory. Union casualties were fifteen killed and nine wounded. The Confederates suffered four killed and six wounded.[2]

Though Southern forces were victorious at Scary Creek, it quickly became obvious to General Wise that his position in the middle Kanawha Valley was untenable without reenforcements and supplies. With troops under General Cox moving up the valley, and additional Union forces under orders to march from the upper Tygart Valley region, by way of Weston, Sutton, and Bulltown, to cut the Confederate lines at Gauley Bridge, Wise knew he would be caught between two jaws of a pincer movement. Wise realized his ill equipped, poorly trained army, would be no match for the army of Cox. He sent a letter to General Lee informing him of the situation, and saying that his legion was almost without ammunition. Wise had no choice but to order a retreat and he set in motion the preparation for what he later called a "retrograde movement."[3]

The retreat began on July 24, and it was a sad time for many of the local residents who had relatives serving in the Confederate Army. Still others were afraid of the Yankees and abandoned their homes. Coalsmouth resident, Victoria Hansord Teays, recorded the events in her diary:

Gauley Bridge before the Civil War. This bridge was built in 1850 and was burned by Confederate troops in 1861. AUTHOR'S COLLECTION

The road was now full indeed of vehicles containing women and children, families going before the army, some with their household goods, others with cows and horses, sheep and dogs. Some lamenting that the valley must be evacuated, some not caring. No one who has not seen the retreat of an army (although in no haste at all) can concieve any idea of it. Towards evening the cavalry and baggage wagons, the artillery near the last. We were all standing out in front of my uncles house, very near the turnpike, to see the last of them. I remember seeing a great many fine looking officers with their brass buttons and waving plumes in their hats. We waved our goodbyes as every woman there was weeping silently. They took off their hats and rode with them in hand until they were entirely past the crowd of weeping women.[4]

The retreat continued all night, some companies not leaving Charleston before the early morning of the 25th. The main body of troops reached Clifton, present Pratt, at two p.m. and camped the rest of the day. Federal forces advanced to Tyler Mountain, a few miles west of Charleston on July 24, and the next morning were met by the Mayor of Charleston and several prominent citizens, who came to surrender the town.[5]

General Cox attempted to ease public concerns that his troops would pillage Charleston, by issuing the following special orders to his command.

> General Orders　　　　　　　　　　　　Hdqrs. District of the Kanawha
> No. 8　　　　　　　　　　　　　　　　　　Elk River, July 25, 1861
>
> The army is about to enter Charleston, which yesterday was the headquarters of Wise and the rebel army. Its people have been told that we come as robbers and murderers of women and children. The General in command knows that every soldier desires to prove that we have been vilely slandered. To make the proof most signal the army will not halt in the town. We will march through in soldierly order, no man leaving the ranks or shouting or making any unnecessary noise. Let the conduct of the troops be in contrast to the profane and disorderly conduct of the rebel army, and the people will bless us as the restorers of safety and liberty of conscience and speech and the defenders of their property. This order will be read at the head of every company before entering the town of Charleston.
>
> By command of J.D. Cox, Brigadier General, commanding district:
>
> 　　　　　　　　　　　　　　　　　　　　　　　　　　J. N. McElroy,
> 　　　　　　　　　　　　　　　　　　　Acting Assistant Adj. General[6]

As the Yankees entered Charleston they experienced first-hand the divided loyalties of its inhabitants. In one place an old man who claimed to have been imprisoned for Union sentiments, was overcome with joy at the sight of Federal troops. He mounted a great rock by the roadside and extemporized a speech in which thanks to the Union army and the lord were curiously intermingled.[7] In another place, two women who were ardent Rebels, claimed not to blame the native born Yankees, but wished that every Southerner in their ranks might be killed.

The Confederate soldiers reached Gauley Bridge July 26, exhausted and hungry. Many of their men were without shoes and others were sick and broken down. Since they were still without tents, some of the men took shelter inside the covered bridge which spanned the Gauley.[8] This bridge had been built in 1850 and was the third bridge spanning the river at that point, the first having been built in 1821. Its large timbers were hand hewn, resting on three piers, the lower one almost in shore. The sides were weatherboarded, with a shingle roof, the interior lighted by four large windows on each side, and the whole painted white.[9]

The situation at this time was very chaotic. The men were attempting to obtain provisions from the post commissary officer but received little attention from him as he was under orders to pack everything as rapidly as possible. General Wise was stamping about the area, "cursing with as much energy and fluency as a trooper in flanders."[10] Many of the Rebel troops had joined the army to defend their homes and were reluctant to leave the Kanawha Valley. Approximately five hundred men and

Zachariah Johnson of the 3rd Rgt. Wise Legion. He enlisted at Gauley Bridge in June 1861.
COURTESY ROBERT BECKELHEIMER, OAK HILL, WV

officers deserted on the march from Charleston to Gauley Bridge. Others wanted to go home and await further orders. Some of the men did manage to slip away long enough to see their families.[11] Beuhring H. Jones of the Dixie Rifles, gave a vivid account of the situation at Gauley Bridge:

> Such demoralization as then ensued has been seldom witnessed. One entire company, perhaps two, deliberately filed off and went home. Another scattered like frightened sheep; but the Captain marched boldly on alone until he encountered a barrel of whiskey; there he halted, got "tight," broke his sword and wore his bars no more, and was never heard of again. Huge sides of bacon were pitched into the mud and trampled under foot. The heads of whiskey and molasses barrels were knocked in, and every man helped himself. The Gauley Bridge that had cost $30,000 was burned although the river was fordible for infantry and cavalry about one hundred yards above. It was said, though I never credited the report, that the famous Hawks Nest was examined with an eye for its destruction, but was declared non-combustible, and was thus saved for the admiration of future tourist. Every man went it on his own hook. For the first twelve hours, despite the efforts of the General, orders were disregarded and system was lacking.[12]

The burning of Gauley Bridge was described by Addison B. Roler of the First Regiment Wise Legion:

> Several men were employed in tarring the side walls of the bridge and arranging the loose material in it so as to burn with the greatest rapidity. A heavy rain in the mean time came up though we kept pretty dry having taken refuge under a ledge of rocks by the road side. The most of our baggage got quite wet and muddy, not as yet having any baggage wagons in which to put them. All the other companies shared the same or a worse fate with their baggage. The road side for half a mile was strewn with tents, frying pans, coffee pots, ham and middlings of meat, and a vast etc. of such like soldier accommodations. The disorder could not have been worse if the main body of the enemy had been right on our heels, while it was at least six or seven miles distant. When all the companies were over the bridge with their baggage and commissary stores, the bridge was set on fire at about 11 p.m. It burned very fast, and the first arch that was fired fell in about one half hour. The whole length of the bridge was at least 150 yards, and ten minutes after the torch was first touched, the whole bridge was one sheet of flame, and for five or ten minutes afterwards presented one of the most beautiful sights I have ever saw. The night was somewhat cloudy and very damp from the recent rain, though it had stopped raining by this time. The smoke arose from above in heavy spiral columns which lingered a moment over the burning wreck, affording time to be lit up in the most gorgeous colors, and then passed off into the air. The curve of the wind was south east and was right against the side of the bridge which caused the smoke from the flooring of the structure to circle beneath the arches in beautiful curves, and to mingle with that of the roofing after it had passed across. The exclamation of all present was what a beautiful sight. I could not but feel for the loss of the property, though I admit of its being a military necessity. Still the sight was enchanting and beautiful that I did wish its continuence for a longer period.[13]

The retreat continued and the Wise Legion reached Bungers Mill, a few miles west of Lewisburg, on July 31. Here, Wise thought it best to refit and reorganize his troops. He also considered a junction between his force and that of General John B. Floyd, then operating in southwest Virginia to protect the line of the Virginia and Tennessee railroad. General W. W. Loring was commanding the Rebel army of northwest Virginia and Wise also considered a junction with him.[14] While at Bungers Mill, an epidemic of measles broke out in the command, further demoralizing the troops. One day General Wise overheard a private speak of "Wise's retreat from the Gauley," to which Wise thundered, "Retreat"— "never dare call it a retreat again sir, it was only a retrograde movement sir." To which the private replied, "I don't know nothin about your retrogrades General, but I do know we did some damn tall walkin."[15]

Wise sent a letter to General Lee detailing the reasons for his withdrawal from the valley and offering his suggestions as to the best course of action. He declared:

> The Kanawha Valley is wholly disaffected and traitorous. It was gone from Charleston to Point Pleasant before I got there. Boone and Cabell are nearly as bad, and the state of things in Braxton, Nicholas, and part of Greenbrier is awful. The militia are nothing for warlike uses here. They are worthless who are true, and there is no telling who is true. You cannot persuade these people that Virginia can or ever will reconquer the northwest, and they are submitting, subdued, and debased. I have fallen back not a minute too soon. And here let me say, we have worked and scouted far and wide and fought well, and marched all the shoes and clothes off our bodies, and find our old arms do not stand service. I implore for some (one thousand) stand of good arms, percussion muskets, sabers, pistols, tents, blankets, shoes, rifles, and powder.[16]

While Wise was at Bungers Mill pondering his predicament, General Cox was at Gauley Bridge basking in his success. The army under Cox reached Gauley Bridge on July 29, just two days after the Rebels withdrew. The Yankees found at Gauley Bridge 1,500 smoothbore muskets and a quantity of ammunition. General Cox fully understood that Gauley Bridge was the gate through which all important movements from eastern into southwestern Virginia must necessarily come, and that it formed an important link in any chain of posts designed to cover the Ohio Valley from invasion. It was also the most advanced single post which could protect the Kanawha Valley.[17]

To protect the camp at Gauley, several outposts were quickly established. The First Kentucky Infantry was placed at a saw mill on the south side of Kanawha Falls. This position was intended to guard the road that came from Fayetteville by way of Big Falls Creek. Two regiments were placed near the bridge, upon the hillside above the hedgerow. Smaller posts were placed a short distance up the Gauley Valley toward Summersville. Cox's headquarters tents were pitched in the dooryard of a dwelling house facing Gauley River, while he occupied an unfinished room in the house for office purposes.[18]

Scouting parties were sent out to study the topography and collect whatever information they could from the local inhabitants. Many of the Ohio troops had never been in the mountains and were thrilled with the natural beauties of the region, as described by Ambrose Bierce, a young soldier from Ohio:

> Nine in ten of us had never seen a mountain, nor a hill as high as a Church spire, until we had crossed the Ohio River. In power upon the emotions, nothing, I think, is comparable to a first sight of mountains. To a member of a plains tribe, born and reared on the flats of Ohio or Indiana, a mountain region was a perpetual miracle. Space seemed to have taken on a new dimension; areas to have not only length and

breadth, but thickness. West Virginia was an enchanted land. How we reveled in its savage beauties. With what pure delight we inhaled its fragrances of spruce and pine.[19]

The hamlet of Gauley bridge consisted of a few dwelling houses, a country store, a tavern, and a church, irregularly scattered along the base of the mountain and facing the road which turned from the Gauley Valley into that of the Kanawha. The slope of the hillside behind the houses was cultivated, and a hedgerow separated the lower fields from the upper pasturage. Above this gentler slope the wooded steeps rose precipately; sandstone rocks jutted out into the crags and walls. In the angle between the Gauley and New Rivers rose Gauley Mount, the base a perpendicular wall of rocks of varying height, with high wooded slopes above. There was just room for the road between the wall of rocks and the water on the New River side. At Kanawha Falls, a saw mill had been built on the south side, and a mill for grinding grain had been built on the north side.[20] General Cox reported that nearly all the population below Gauley bridge was loyal to the Union, while the population above were mostly secessionists. Another officer wrote that "although the inhabitants were not ready to do much for the rebel cause, they would do less for ours."[21]

Previously unpublished photo of Gauley Bridge taken during the Civil War. COURTESY AUBREY MUSICK, GAULEY BRIDGE, WV

During the first week of August General Cox established parade grounds and other training areas for his troops. Scouting parties that had been sent as far into the country as Sewell Mountain reported no organized Confederate force was then in the area. On July 23 General William S. Rosecrans had been given command of the Department of the Ohio, embracing a portion of Western Virginia. General McClellan had been ordered to Washington after the Union defeat at Manassas, Virginia, on July 21. On August 7 Cox sent a letter to General Rosecrans who was then at Clarksburg with his command:

> Headquarters Kanawha Brigade,
> Gauley Bridge, August 7, 1861
>
> Brig Gen. W. S. Rosecrans,
> Comdg. Army of Occupation, Western Virginia, Clarksburg, Va:
>
> General: . . . Since arriving here I have had reconnoitering parties under intelligent officers at Fayette Court House, Sewell Mountain, Summersville, and intermediate points, the substance of whose information is contained in the reports accompanying this. The retreat of Wise has every characteristic of a final movement out of the valley. Not only his burning of bridges and destruction of arms and other property has this look, but the conduct and air of the professed secessionist strongly confirms this opinion. Great numbers of Wises troops, raised here in the valley, deserted him near this point, and the story of the deserters is quite uniform that it was understood that he was permanently abandoning the valley. . . . The town of Charleston is the headquarters of the secessionist in this valley, and I have kept a regiment there since I came through the place. . . .
>
> J.D. Cox
> Brigadier General, Cmdg.[22]

To further strengthen the post at Gauley Bridge, General Cox had constructed an epaulment for cannon high up on the hillside, covering the ferry and the road up New River. An infantry trench with parapet of barrels filled with earth, was run along the margin of Gauley Bridge till it reached a creek coming down from the hills on the left. There a redoubt for two guns was made, commanding a stretch of road above, and the infantry trench followed the line of the creek up to a gorge in the hill. On the side of Gauley Mount, facing the camp, timber was slashed from the edge of the precipice nearly to the top of the mountain, making an entanglement through which it was impossible that any body of troops could move. Below the falls the saw mill was strengthened with logs until it became a block-house loopholed for musketry, commanding the road to Charleston, the ferry, and the opening of the road to Fayette Court House.[23]

FIELD AND STAFF OFFICERS OF WISES LEGION
AUGUST 1861

Brig Genl/ Henry A. Wise
Asst Adjt Genl/ Capt. Wm. B. Tabb
Aid de Camp/ W. Bacon
Asst QM of Brigade/ Capt. F. D. Cleary
Asst Commissary/ Capt. Wm. H. Thomas

CORP OF ENGINEERS
Chief/ Capt. Bolton
    Capt. T. T. L. Snead
    1st. Lt. George Bagwell
    1st. Lt. Arch Blair
    2nd. Lt. S. A. M. Syme
Chief of Ordnance for Brigade/ Capt. L. Buckholtz
Medical Director/ Surgeon A. O. Crenshaw

CAVALRY REGIMENT
Colonel/ J. Lucius Davis
Lt. Col./ John N. Clarkson
Major/ C. B. Duffield
Adjutant/ M. J. Dimmock
Asst. QM./ S. C. Ludington
Asst. Commissary/ J. H. Vandiver

CORPS OF ARTILLERY
Lt. Col./ W. H. Gibbs
Adjutant/ C. Ellis Munford
Asst. QM./ L. N. Webb
Asst. Commissary/ John Mason

1st. REGIMENT INFANTRY
Colonel/ John H. Richardson
Lt. Col./ Nat Tyler
Major/ H. W. Fry
Adjutant/ Lt. Henry A. Wise
Asst. QM./ N. S. Thomas
Asst. Commissary/ A. Kinney

2nd. REGIMENT INFANTRY
Colonel/ C. F. Henningsen
Lt. Col./ F. P. Anderson
Major/ John Lawson
Adjutant/ Lt. John R. Blocker
Asst. QM./ J. C. Deane
Asst. Commissary/ A. W. Matthews

3rd. REGIMENT INFANTRY
Lt. Col./ James W. Spalding
Acting Major/ Capt. W. A. Swank
Adjutant/ Lt. J. H. Pearce
Asst. QM./ Joseph M. Brown
Asst. Commissary/ Huston Estill

General John B. Floyd, C.S.A. He was a "political general" having served as governor of Virginia before the war. His lack of military experience contributed to the South's defeat in Fayette County.
COURTESY MUSEUM OF THE CONFEDERACY

■ *Chapter Three*

The Scene of Action

On August 6 a meeting was held between General Wise and General John B. Floyd, at White Sulphur Springs, to which point Wise had moved his command after remaining briefly at Bungers Mill. Wise and Floyd were both former governors of Virginia, and as rival politicians had developed a hatred for each other which they brought with them into the war. By virtue of his earlier commission Floyd was senior to Wise and thus entitled to command their combined forces when they were operating together.[1]

Their meeting at the White Sulphur was anything but amiable. General Floyd, as holding superior rank, received with patronizing air General Wise's address, and the report as to the state of his Legion.[2] Wise was known for long "windbag" speeches, and he took the opportunity of this meeting to exercise his considerable talents. He stood up, placed his hands on the back of his chair, and spoke for almost two hours. He reviewed the history of the United States from its discovery, the Revolutionary War, the Mexican War, the causes which led to the current troubles, his march down the Kanawha River, the affair at Scary Creek, and his retreat to the White Sulphur.[3] General Floyd was not impressed by the speech, and he told Wise that it was his intention to retake the Kanawha Valley as soon as possible. Wise was firmly against the move, saying it would take at least two weeks to refit his command and obtain the wagons needed for an advance.[4]

Despite Wise's objections, Floyd informed him on August 8 that he was anxious to begin a move toward the Kanawha Valley. Floyd asked for a detailed list of the forces Wise had available for duty. In his response, Wise displayed his contempt and unwillingness to cooperate with Floyd. He stated that at no time had his men been supplied with sufficient clothing, camp equipage, arms, tents, shoes, and other needs and that he was also greatly in need of wagons and that 300 of his men were in the hospital.[5] Actually, General Floyd's command was in no better condition, and fully half of his men were sick with measles.[6]

On August 10, Wise wrote to General Lee, who was then camped at Valley Mountain, in Pocahontas County.[7] Among other things, Wise asked if he might be sent some weapons that had been taken from the Yankees at Manassas. He had written to Lee previous to that in an unsuccessful attempt to have his command permanently separated from that of Floyd.[8] The next day, August 11, Floyd assumed total command of all Confederate forces intended to operate against the Yankees in the Kanawha Valley and adjacent country. This order did not set well with Wise, and there began what was to become a death blow to Confederate efforts in Fayette County and southern West Virginia generally.[9]

General Wise wanted to draw the Union army into the eastern wilderness of Fayette County. He argued that to advance all the way to the Kanawha River would force their army to haul supplies an additional forty miles over some of the worst roads in western Virginia, roads that had been turned into a sea of mud by almost daily rainfall. Despite Wise's plan, Floyd wanted to advance to the Kanawha Valley and drive General Cox back across the Ohio River. General Lee was able to temporarily settle the matter by choosing Wise's plan as the best. This little victory for Wise was not enough to ease his jealousy of Floyd.[10]

On August 13 Wise wrote to General Lee in another attempt to separate his command from Floyd's, and asked that Floyd's orders to his brigade be communicated through him. Lee responded that he felt no orders were necessary on the subject, and attempted to pacify Wise's ego: "I hope," he wrote, "I need not assure you that I never entertained the least doubt as to your zealous and cordial cooperation in every effort against the common enemy. Your whole life guarantees the belief that your every thought and act will be devoted to the sacred cause, dearer than life itself, of defending the honor and integrity of the State."[11]

Unfortunately, cordial cooperation would never occur between Wise and Floyd. Had they acted together at this time, their total force, when combined with the militia of Beckley and Chapman, would have exceeded 6,000 troops. Acting together, they could have made matters very difficult for the 5,000 troops then under command of General Cox.

On August 14, after growing impatient with the delaying tactics of Wise, General Floyd peremptorily ordered Wise to march with all the forces under his command, to join him at his camp near Meadow Bluff, Greenbrier County.[12] Wise complied with this order on August 15, after writing a letter of complaint to General Lee, advising him of the situation and offering his opinions as to the best course of action.[13] As Wise was on the march, General Floyd sent a scouting party to Sewell Mountain, Fayette County, and began moving his entire command to that point. The scouts, under command of Colonel Heth and Colonel J. L. Davis, clashed on the western foot of Big Sewell Mountain, with a 120 man scouting party from the 11th Ohio Infantry, commanded by Colonel Joseph W. Frizell. Observing the Rebel cavalry advancing up the mountain, the Yankees had deployed their force into the thickets on the left side of the turnpike. As the advance party of Rebels drew near they caught sight of the Yankees and opened fire. A brisk skirmish was kept up for several minutes with losses being three Yankees wounded and five Confederates killed and wounded. The Federals withdrew a short

Colonel Joseph W. Frizell—11th Ohio Volunteer Infantry. His men ambushed Confederate cavalry at Sewell Mountain and Hawks Nest.
FROM THE 11TH REGIMENT O.V.I. BY HORTON & TEVERBAUGH, 1866

distance and formed an ambush along the road. Realizing the Rebels were not going to fall into their trap, the order was given to fallback to their camp at Locust Lanes, four miles west of Sewell Mountain.[14] While this first clash of arms at Sewell was a minor affair, it served as a barometer of events which would transpire in the coming weeks.

Cox realized that it was just a matter of time until his army and the army of Floyd would clash in more than just a brief skirmish. From his post at Gauley Bridge he wrote to General Rosecrans at Clarksburg asking for a quantity of ammunition.

Gauley Bridge, August 15, 1861

General Rosecrans,
    Clarksburg, Va:

I do not learn of any great change in the enemy's position since yesterday. We have about forty five rounds per man of musket and rifle cartridges, and about 140 rounds for each of our five cannon, two of which are rifled. I want at least 200,000 musket cartridges; caliber .69; 20,000 enfield cartridges, caliber .57; 10,000 ditto, caliber .58; 50,000 cartridges for the Greenwood altered rifle, bright barrels; 200 rounds each James solid shot and James shells, and 300 each of grape and canister for smoothbore, with 150 solid shot, all for six pounders. Extra caps for muskets, 30,000; also 1,000 friction primers for cannon. If a persistent defense is to be made here we should want much more than the above, and it should be where we could easily get it.

J.D. Cox
Brigadier General, Commanding[15]

On August 16 Floyd and Wise arrived in the vicinity of Sewell Mountain, with Wise camping near the eastern foot. The next day, Floyd advanced with a portion of his command to the western foot of Big Sewell Mountain and established headquarters at the home of Frank and Margaret Tyree. This house still stands and is known as the Old Stone House, or Tyree Tavern. It is located along the old James River and Kanawha Turnpike, near Ravens Eye. This house was built in 1824 by Richard Tyree, who had moved to Fayette County from Richmond in 1816. Used as a stagecoach stop and inn, the house became a prized possession by the commanders of units operating in the vicinity.

Besides being used as a headquarters by Blue and Gray, it was also used as a military hospital.[16] There are many old stories surrounding this house, one of which tells of how the blood of wounded soldiers dripped on the floor and soaked into the boards. The original flooring was replaced in 1920 and during the replacement several cannon balls were found under the house, as well as a few discarded medical instruments. Many famous people have stopped at the Stone House, even Andrew Jackson stopped there for a few days to hunt during October

1832. John Robinson who was a partner of Horace Greely in his newspaper work was a frequent visitor there. Matthew Fountain Maurry, known as the "Pathfinder of the Seas," drew the charts for "Steam Ship Lanes" at the old Stone House while recovering from an accident from the wreck of the stage coach. The Stone House also had a saloon accommodation, just across the road where most of the men spent their leisure hours. The saloon was later known as the "Red Rabbit." The servants' quarters were located on the hill to the right of the trail going east; the building was later demolished by fire. In the earlier years the Old Stone House was very considerate in the price of their accommodations. A bed could be had with or without sheets for the small difference of two cents. Imagine getting your choice of a bed for 5 or 7 cents! The sleeping arrangements were very peculiar to what one might expect nowadays. Ladies must sleep in the east side of the house while the men must sleep on the west side. The house remained in the hands of the Tyree family until 1884 when it was purchased by the Longdale Iron Company, which began the development of its coal lands at Cliff Top, two miles west of the Tyree home. In 1963 it was purchased by the Reverand Shirley Donnelly, who set about putting the building in good condition to preserve it as a noted state landmark. Today the home is privately owned and is in excellent condition.

The Old Stone House, or Tyree Tavern, built in 1824. COURTESY STAN COHEN

On August 20, the commands of Floyd and Wise left Sewell Mountain, headed for the Kanawha Valley.[17] Being aware that enemy scouts were probably in the area, advance parties of cavalry were ordered out ahead of the main column. The Rebel cavalry split up at Dogwood Gap, near present day Hico. Some of the men were sent down the Sunday Road while the others, commanded by Colonel Croghan, proceeded along the turnpike towards Hawks Nest. It didn't take long for skirmishes to develop between Confederate and Union scouting parties. On the Sunday Road a small group of Yankees from the 7th Ohio Infantry clashed with a superior force of the Wise cavalry. This fight was severe but of short duration. Captain John F. Schutte, Company K, of the 7th Ohio, was killed, and thus became the first Union army officer to be killed in Fayette County during the war. Sergeant Edward H. Bohm, of the 7th Ohio, later wrote of the incident: "I was running along side Cpt. Schutte in the middle of the road, bullets zipping all around us and about us. My canteen fell, its string cut by a bullet; my cap fell off my head a little ahead of me, with a bullet through it. Poor old Private Charles Rich, to the right, a little ahead of me, dropped with a yell of pain and crawled into a fence corner. All at once Cpt. Schutte groaned, 'I am shot.' Stopping a moment I saw a bullet hole back and front."[18] Indeed, a rifle ball had struck Schutte to the right of his spine, coming out a little to the left of his navel.

Captain John F. Schutte of Company K—7th Ohio Infantry. He was the first Federal officer killed in Fayette County during the Civil War. COURTESY TERRY LOWRY, SOUTH CHARLESTON, WV

On the turnpike members of the 11th Ohio Infantry had constructed breastworks in the middle of the road near Hawks Nest, and east of Hawks Nest near Piggotts Mill. Companies A and H, under command of Major Coleman, were posted along a fence in the edge of the woods; Company C behind the breastworks in the road; and the balance of the regiment about one half mile to the rear, behind additional breastworks.[19] Assisting the 11th Ohio were two "Snake Hunters," from an independent group of loyal partisan rangers. These rangers were from a select group of men who had been assembled by Captain John P. Baggs. They were said to be stalwart, rugged foresters, shrewd—wary, and daring. Their business was to "trot" in the extreme front, in the capacity of guides, scouts, and spies. They wore no particular uniform, and were allowed to indulge their fancy as to the choice of arms.[20]

As the Rebel cavalry under Croghan drew near, the "Snake Hunters" opened fire. Taken by surprise, the Confederates returned fire but were thrown into confusion. Just as the Yankees had been outnumbered in the fight on Sunday Road, the Rebels at Hawks Nest were outnumbered five to one. The Rebels were forced to withdraw, leaving behind one killed and three wounded. The Yankees suffered two killed and two taken prisoner.[21] In the fighting on Sunday Road the Union forces lost two killed, five wounded, and five captured.[22]

One of the Confederate soldiers wounded at Hawks Nest was Private Steward D. Painter, of Wythville, Virginia; he was thirty four years old, and was a member of Company B, 45th Va. Infantry. Painter had been shot through his left lung and was carried to the home of Matilda Hamilton, near present day Hawks Nest lodge. Two Federal surgeons, Dr. Gill and Dr. A. B. Hartman, nursed Painter back to health. Dr. Gill kept Painter's shotgun and found that one barrel had been discharged and the other contained fifteen buckshot. In less than two weeks Painter became strong enough to be passed back into Confederate lines so that he might go home and complete his recovery. Before he left he sent Dr. Gill a letter by flag of truce:

Mrs. Hamiltons, September 4, 1861
Dr. Gill, Dear Sir

> I write this for the purpose of expressing in written language my gratitude to you for the generous, kindly treatment you bestowed on me, who your enemy, rendered unfortunate by the fate of war, was thrown upon your mercy. Sir, it is impossible for me to express all that my heart dictates. Suffice it to say I can never forget you. No matter what may be the period of my life or the circumstances that may surround me, whether in peace or war, prosperity or adversity, the remembrance of Dr. Gill will abide with me ever, and toward him will flow unceasingly my hearts deepest gratitude. May heaven smile upon you Doctor; may your path be strewn with lifes choicest flowers; may you pass unscathed

War period view of Hawks Nest, looking east. COURTESY SWV

Federal pickets posted at the "Hawks Nest" Fayette County. COURTESY SWV

through the horrors of this unnatural war, and when you die may these words be your stay and suport—"inasmuch as ye did it unto the least of these, ye did it unto me." Gen. Wise and the officers of his staff wish me to convey to you their respectful regards.

Ever your friend,
Steward D. Painter[23]

The skirmishes of August 20 caused General Cox to become concerned that a superior force might attack his position at Gauley Bridge. Accordingly, he ordered the 7th Ohio Infantry, under command of Colonel Erastus B. Tyler, to leave their camp at Kesslers Cross Lanes, near Carnifex Ferry, and proceed to Twenty Mile Creek. With this movement General Floyd realized that the way was now clear for him to occupy Carnifex Ferry, and eliminate an important link in the Union Army's chain of communications between the Kanawha River and the forces to the northwest under Rosecrans.

On the night of August 21 Floyd crossed the Gauley River at Carnifex Ferry. Once on the north bluffs overlooking the Gauley River, in a horseshoe-shaped bend of the river, Floyd began to entrench, designating his location as Camp Gauley.[24] The next day he wrote to General Lee informing him of his activities, and asking for three regiments to replace the Wise Legion.[25] The animosity and lack of cooperation between Floyd and Wise were greatly hampering Confederate efforts in the region, and both men acknowledged it.

On August 24 General Wise wrote to Lee from his camp near present-day Hico. General Wise complained about what he perceived to be Floyd's attempts to merge their two commands, and he asked Lee once again to separate their armies. Wise wrote:

> To be plain, Sir, I am compelled to inform you expressly that every order I have received from General Floyd indicates a purpose to merge my command in his own and to destroy the distinct organization of my Legion. We are now brought into a critical position by the vacillation of orders and confusion of command. . . .
>
> I now ask to be entirely detached from all union with General Floyds command. I beg you, sir, to present this request to the President and Secretary of War for me. I am willing, anxious, to do and suffer anything for the cause I serve, but I cannot consent to be even subordinately responsible for General Floyds command, nor can I consent to command in dishonor. I have not been treated with respect by General Floyd, and cooperation with him will be difficult and disagreeable, if not impossible. I earnestly ask that while he is attempting to penetrate Gauley I may be allowed to operate in separate command from him, but siding his operations, by being ordered to penetrate the Kanawha Valley on the south side, by the Loop or Paint Creek, or by the Coal River; or send me anywhere, so I am from under the orders of General Floyd.[26]

Due to military necessity, General Wise did not get his wish at this time. Lee responded: "The Army of the Kanawha is too small for active and successful operations to be divided at present. I beg, therefore, for the sake of the cause you have so much at heart, you will permit no division of sentiment or action to disturb its harmony or arrest its efficiency." General Wise was undoubtedly frustrated by his failed attempts to separate his command from Floyd's.[27]

On the morning of Sunday, August 25, the lack of cooperation between these two men resulted in a demoralizing defeat of Confederate cavalry. Acting Colonel Albert Gallatin Jenkins advanced from Floyd's camp at Carnifex Ferry into the lines held by the Wise Legion near Hawks Nest. With Jenkins was Major Reynolds and 175 men of Floyd's cavalry. On reaching Piggots Mill, a short distance east of Hawks Nest, Jenkins relieved the eighteen-man picket force which had been posted there by General Wise and was well acquainted with the area.[28] Knowing that the Union lines were nearby, the pickets tried to warn Floyd's cavalry against an advance. Ignoring their pleas, Jenkins and his men rode directly into an ambush which had been set up by Lieutenant Colonel Frizell of the 11th Ohio Infantry. Frizell's men were hiding in the woods on both sides of the turnpike near the Hawks Nest.[29] As the Rebel cavalry advanced along the road they clashed with a small party of skirmishers which Frizell had sent out to draw the Confederates into the ambush. When Frizell's men retreated, Jenkins's cavalry continued their advance and were allowed to pass between the Yankees who were secreted in the woods. When Jenkins's column was about half surrounded the Yankees opened fire. Instantly, several Confederates fell from their horses and Colonel Jenkins's horse was shot under him. A complete state of panic came over the Rebels and many of them threw down their guns, knapsacks, canteens, and anything that would impede their flight. Some of the men managed to return fire but they were so completely surprised and surrounded they had no chance of success.

Captain John P. Brock, who was in charge of the picket force Jenkins had relieved, came rushing down the road with twenty of his men. Finding himself in no better situation than Jenkins, his men exchanged a few shots with the enemy and joined the others in a retreat. The first of Floyd's cavalry to escape the ambush fled with all the speed they could muster to General Wise's camp at Dogwood Gap, five miles to the east. Within fifteen minutes Wise started eighteen companies of infantry and three pieces of artillery on a double quick march to Piggots Mill.[30] Before the reenforcements could arrive Jenkins's force suffered a humiliating defeat. Sixteen Confederates were wounded, including Colonel Jenkins, who was badly bruised in the fall from his horse. One Rebel was killed and two of the wounded were taken prisoner. They also lost three horses, twenty hats, two saddles, and miscellaneous equip-

Colonel Albert Gallatin Jenkins. He fought under General Floyd in the Kanawha Valley, led a raid through central West Virginia in 1862, and was killed at the battle of Cloyd's Mountain in 1864.
COURTESY NATIONAL ARCHIVES

ment that had been left strewn along the road. Federal losses were comparatively light with only a few wounded.

This fiasco with Floyd's cavalry infuriated General Wise. He sent Floyd an angry letter of complaint, and sent General Lee a multipage letter of grievance against Floyd.[31] Major Thomas L. Broun, of the Third Regiment, Wise Brigade, was among the men sent to reenforce Jenkins in his fight. The next day Major Broun wrote a letter to his wife describing his experience:

> Camp Dogwood, near Hawks Nest, Fayette Co. Va.
> August 26, 1861
>
> My Dear Annie:
>
> Recently Joe and I have received from you and Len several letters, for which you have our sincere thanks. A soldiers life in Western Va. is a hard fate. Sometimes we march all day without eating anything, and then sleep at night on the wet ground. Our forces number some 4,000 men in this neighborhood. A good many were left behind at the White Sulphur Springs, sick with measles. The enemy are stationed at Hawks Nest, eight miles from us. On yesterday (sunday) we thought we would have a big fight. About 1000 of our forces were ordered out. We marched right upon the enemy but they retreated. They had cut to pieces in a terrible manner about 150 of our Cavalry within five miles of us. In marching upon them, I could not, Annie, but reflect upon our condition as we marched hurriedly, on not knowing but many of us might fall in the expected battle. Strange did it appear to see men on a beautiful, bright sabbath (Gods Holy Day) armed to the teeth and rushing hurriedly on into the jaws of death. Yes, and the country around was looking lovely, the mountain scenery beautiful and picturesque. Everything seemed grateful to providence, the cattle in the field, the birds in the forest, the green grass, and the rich foliage—but man, of all Gods creatures and created things, seemed to be maddened and bent on his own ruin and destruction. Feeling however, perfectly conscious of the justice of our cause, and that we had right on our side, I marched along with the column, feeling as self possessed, calm, and collected as I do now, every now and then meeting a poor horse or man covered with fresh blood, and some dying and suffering agonies from their wounds. War certainly has a domoralizing effect upon many soldiers and officers. It has impressed me however, with the utter vanity and fickleness of all earthly things. It has doubly assured me that all these trials and troubles are sent upon us for our ungratefulness to the maker of all things. I, therefore, patiently endure all the trials and perplexities incumbent on this terrible war. I have resolved to do my duty and trust in providence for the issue, believing that this is all a poor mortal can do. Write frequently. We are delighted to hear from home. Your letters are a great joy to both Joe and myself. I am very proud that our family are all so heartily engaged in this war. May God grant a speedy termination.
>
> Love to all,
Affectionately yours,
Thos. L. Broun[32]

Thomas Lee Broun of the 3rd Rgt. Wise Legion.
COURTESY SWV

Gauley Bridge as it appeared in 1861. Troops of General W.S. Rosecrans camped on the hills above the bridge.
COURTESY SWV

Previously unpublished photo of Hawks Nest taken during the Civil War. This view is looking east. COURTESY AUBREY MUSICK, GAULEY BRIDGE, WV

Fortunately for General Floyd, he never received the bad publicity which he certainly deserved after the defeat of his cavalry. During the late afternoon of August 25, Colonel Erastus B. Tyler returned with his 7th Regiment, Ohio Infantry, to Cross Lanes, Nicholas County. This position was less than three miles from the Rebels at Camp Gauley. General Floyd's cavalry reported Tyler's return to the area and the decision to attack was made.

Colonel Tyler failed to post sufficient pickets or properly scout the area. His nonchalant attitude contributed to his defeat. Just after dawn of August 26, the Confederates surprised Tyler's men while they were eating breakfast. The Ohioans were outnumbered nearly three to one, but after initially being routed, they managed to form a defense and fought for nearly thirty minutes before being compelled to retreat. When the smoke had cleared the victorious Confederates reported at least two Yankees killed, twenty-nine wounded, and 110 captured. The Rebels lost one regimental flag and had several men killed and wounded.[33]

Needless to say, General Floyd hailed this as a great victory. Considering that the Rebels had vastly outnumbered the Ohioans, their victory should have been more complete. Nevertheless, they managed to temporarily sever an important link in the Union army's line of communication. General Wise was jealous of Floyd's success, and he referred to the fight as the "Battle of Knives and Forks" (because the Yankees were eating breakfast when attacked).

After this victory General Floyd intensified his entrenching efforts at Camp Gauley and pondered an advance into the Kanawha Valley. Wise was with his command at Dogwood Gap, some 1,436 strong.[34] When news of Tyler's defeat reached General McClellan he was furious. He ordered General Rosecrans to proceed south from Clarksburg and attack Floyd. This movement by Rosecrans into Nicholas County set the stage for the battle of Carnifex Ferry.

Adding to the many worries confronting General Cox at Gauley Bridge was the occupation of Fayetteville by the 142nd and 184th Virginia Militia, under Colonel Beckley. Receiving information from scouts and Union citizens that Confederate forces were massing south of New River, Cox advanced his pickets further south on the Fayette road at Cotton Hill. Colonel Beckley's cavalry scouted the Fayette road towards Kanawha Falls, and this resulted in a skirmish on August 28. In this fight Cox lost one man killed and two wounded; the Rebels had no casualties. Wise received reliable information as to the strength of the Federal post at Gauley Bridge, and he began planning an attack on that position.[35]

Problems continued to mount for General Cox when on August 29, there was a mutiny by some members of the 2nd Kentucky Infantry. Lieutenant Wagner of the topographical engineers discovered a branch

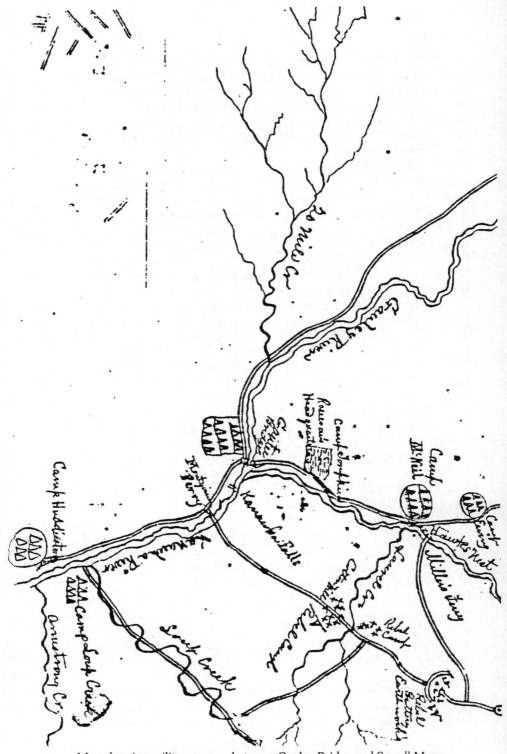

Map showing military camps between Gauley Bridge and Sewell Mountain. Drawn by R.D. Van Duerson of the 12th O.V.I.
COURTESY OHIO HISTORICAL SOCIETY

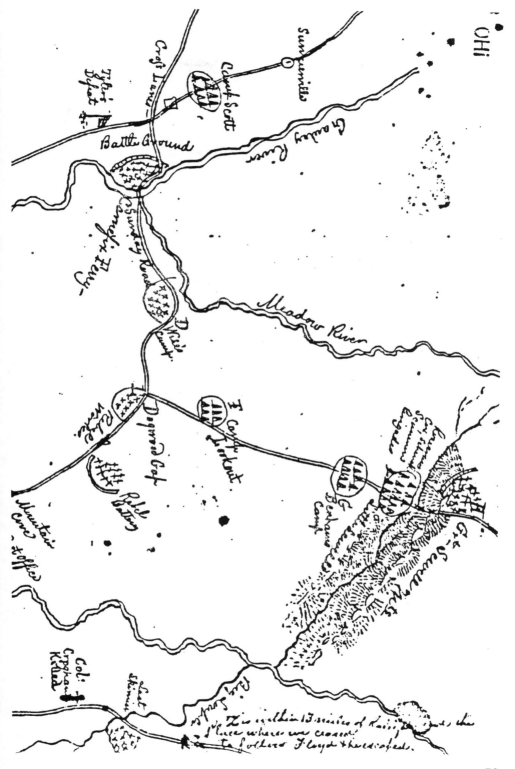

road leading into the valley of Scrabble Creek behind the main camp. It was decided to construct additional trenches along the hills overlooking Scrabble Creek so as to better defend their main camp against attack. Some members of Company F, 2nd Kentucky Infantry, were selected to construct the trenches and marched to the designated area. Sergeant John Joyce became very abusive and violent. He and the other men complained that they were already overworked and that the weather conditions were not right for such duty. Lieutenant Gibbs, also of the 2nd Kentucky, threatened to shoot Sergeant Joyce if he did not resume his place in line.

The matter was now growing increasingly out of hand, with the men yelling in support of their sergeant. Lieutenant Conine, an aide to General Cox, stepped forward and called for guards to come and place Sergeant Joyce under arrest. The argument between the Sergeant and Lieutenant Gibbs continued and before the guards could arrive, Joyce challenged Gibbs to go on and shoot, which he did, killing the Sergeant instantly.[36] General Cox could hear the commotion and he came out of his office to investigate, as he later described:

> I was setting within the house at my desk, busy, when the first thing which attracted my attention was the call for the guard and the shot. I ran out, not stopping for arms and saw some of the men running off shouting, 'Go for your guns, kill him, kill him.' I stopped part of the men, ordered them to take the Sergeant quickly to the hospital, thinking he might not be dead. I then ordered Gibbs in an arrest till an investigation could be made, and ran at speed to a gap in the hedge which opened into the regimental camp. It was not a moment too soon. The men with their muskets were alredy clustering in the path, threatening vengeance on Mr. Gibbs. I ordered them to halt and return to their quarters. Carried away with excitement, they leveled their muskets at me and bade me get out of their way or they would shoot me. I managed to keep cool, said the affair would be investigated, that Gibbs was already under arrest, but they must go back to their quarters. The parley lasted long enough to bring some of their officers near. I ordered them to come by my side, and then to take command of their men and march them away. The sequel to the affair was not reached till some weeks later when General Rosecrans ordered a court martial at my request. Lt. Gibbs was tried and acquitted on the plain evidence that the man was in the act of mutiny at the time. The court was a notable one, as its judge advocate was Major R. B. Hayes, of the Twenty Third Ohio, afterwards President of the United States, and one of its members was Lt. Col. Stanley Matthews, of the same regiment, afterwards one of the justices of the Supreme Court.[37]

General Cox knew that Colonel Beckley's militia was at Fayetteville, but he did not yet know that Beckley had been reenforced by General A. A. Chapman's Nineteenth Brigade, Virginia Militia, on September 1. Chapman's force was 1,500 strong and combined with Beckley's, they presented a formidable army of approximately 2,100 men, including cavalry.[38]

By this time General Floyd had received reports that a large Union force was concentrating at Sutton, and he suspected they were planning to attack his position at Camp Gauley. Being notified on September 2 that Chapman and Beckley had united at Fayetteville, he sent instructions to General Chapman:

> I have received information, which is reliable, that a considerable force of the enemy have collected in Sutton. Whether it be their plan to attempt to reenforce Cox at Gauley Bridge or to unite with him in an advance upon me from that point I have not been able to ascertain. I shall watch their movements with all vigilance and shape my action somewhat accordingly. However this may be, you can render most essential service by pushing your forces on to Charleston and embarrassing, checking, and destroying, if you can, the navigation of the Kanawha, thus holding Cox in his present position at Gauley or thereabouts, or drawn off to Charleston.[39]

General Wise had similar aspirations, and during the late afternoon of September 2 he advanced with 1,250 men from Dogwood Gap to Hawks Nest. This advance resulted in a skirmish with members of the Eleventh Ohio Volunteer Infantry. In his report of the incident General Wise stated:

> Late in the evening I advanced upon Turkey Creek, leading the advance guard myself in person. About dusk we arrived at McGraws bridge, over Honey Creek, and were fired upon (a very short time hotly) by the enemy, concealed in the cornfields and brush woods on both sides, and just as we were crossing the bridge. I am proud to say that the guard (Captain Summers Co.) stood their ground and behaved handsomely, returning the fire promptly, and I led them across the bridge, the enemy disappearing before us on the quick advance of our column. Night coming on, I thought it prudent to rest on our arms for the time.[40]

At this time Wise did not realize that his advance had halted a mere 500 yards from the entrenched position of the 11th Ohio Infantry, at Big Creek, present-day Chimney Corner. In fact, the two lines were so close that one of the Yankee picket posts was actually situated behind the Rebel line, as described by a member of the 11th Ohio: "During the night of September second, Wise advanced and stationed his troops just across a ravine in front of our line. One of our picket stations was so situated that the Rebels had got between it and the regiment before the men could escape, and Alex Garmack and John Helmer, of Company A, laid inside the Rebel lines until near morning, when they succeeded in getting away, having crawled a long distance through the thickets along the road."[41]

Early the next morning a combined attack was made against the Yankees. General Wise attacked at Big Creek with 1,250 men, while Gen-

Luther Warner house which stood at Beckwith, WV. Built in 1856 it was destroyed by fire in 1979. During the Civil War it was used as a hospital and headquarters by North and South.
COURTESY MR. AND MRS. ALLEN HESS, FAYETTEVILLE, WV

eral Chapman attacked along the Cotton Hill road towards Kanawha Falls, with 2,100 militia. The fighting at Big Creek was sometimes heavy, though little progress was made by the Rebels. Shortly after the Confederate artillery had opened the contest, General Cox reinforced the position at Big Creek with half of the 26th Ohio Infantry. This swelled their force to approximately 1,500. While General Cox could easily have added more men to that position, he became concerned that these concerted attacks would be joined by the brigade of General Floyd, attacking his position from the Gauley road. The skirmishing was kept up all day with Chapman's militia succeeding in driving the Union pickets along the Fayette road to near the Kanawha River. At dusk General Wise broke off the attack and fell back to Hamilton's near Hawks Nest, and to Westlakes Creek, in Ansted.[42]

The Federals suffered a total of six men wounded and three captured in the combined attacks. Confederate losses were eleven men killed and wounded. Among the killed was Captain Nathan Louborough, adjutant for General Beckley's 27th Brigade, Virginia Militia. Louborough was the first Confederate officer to be killed in Fayette County during the war. His death occurred during an advanced scouting mission along the Fayetteville road at Cotton Hill.[43] Captain Ralph Hunt, of the 2nd Kentucky Infantry (US) shot Captain Louborough and was then captured along with two of his men. After several months in confinement Hunt was exchanged. Shortly after his release a story about Louborough's death appeared in the National Intelligencer as follows:

> A few men had been sent in advance as scouts, but it seems that these were bewildered among the dwarf pines and bushes, and, in making their way back, unfortunately got into Captain Hunts rear. The Captain, after posting his men, had gone forward a few yards, accompanied by two of his men, and hearing an advance upon the road, stepped forward a few paces, in expectation of seeing his returning scouts, but the party advancing along the road turned out to be the leading files of the advance guard of rebel forces. With these was a fine looking officer named Louborough, who had been sent out to drill the confederate troops in that region. This officer was marching some distance in advance of his men, and catching sight of Cpt. Hunt, poured forth a torrent of imprecations, exclaiming, 'Come out, you damned Yankee son of a bitch, and be shot,' at the same time raising to his shoulder his Mississippi rifle. The Captain had a musket with him, which he instantly leveled at his adversary. The combatants were about fifty yards apart; each fired at the same instant; the adjutants ball whistled close by Cpt. Hunts ear, but the adjutant himself, with a curse upon his lips, fell dead with a bullet through his brain. So instantaneous was the death that not a limb stirred after the body touched the Earth. Not less than seven shots were instantly fired at Cpt. Hunt, none of them taking effect. The enemy, enraged at the loss of a favorite officer, were at first inclined to be revengful, but the gallantry he had just displayed, and the coolness with which he bore himself when in their power, finally won their respect. The men of Cpt. Hunts company supposed their leader to be killed, and made good their escape to camp. Hunt and the two men with him were so surrounded that their escape was impossible.[44]

Captain Louborough's death was indeed a sad event for the Confederates. He had been liked by all, and was a dedicated, enthusiastic officer. When General Floyd received word of his death, he offered his condolences to General Beckley: "I take pleasure in congratulating yourself and General Chapman upon your success in repulsing the enemy in your skirmish on last Tuesday, and upon the eligible position thereby won. I am, however, pained to learn of the death of your adjutant, Captain Louborough."[45]

The day after Louborough's death General Beckley made arrangements for his burial.

Camp Laurel Fayette Sept. 4 1861
Brigade Orders:

    1st. The General commanding announces what has already excited the warm sympathys of both regiments constituting his command. The death of Captain and Adjutant Nathan Louborough at the hands of the enemy on yesterday. While scouting Cotton Hill, he being gallantly in the advance his loss is a great one to the state service and personally to the General commanding almost irrepairable. Our deceased brother died a glorious death, defending the liberties of his native state, and his memory is precious to every true Virginian.

    2nd. Capt. Alex Bryon will have a company of fifty men detailed half from the 142nd and half from the 184th as a funeral escort to the deceased. The Captain will command it. He will be advised of the hour when he will march the escort to Mr. Warners residence.

    3rd. The Adjutant of the 142nd regiment will cause the order to be read to the two regiments, who will be paraded under arms for that object.

    4th. Cpt. Herndon will detach his whole troop as a funeral escort dismounted to the bodies of Privates Frazier and C. R. Hale.

Alfred Beckley
Brig. Genl. 27th Brigade & Col. 35 Rgt.[46]

General Lee had hoped that the combined forces of Wise, Chapman, and Beckley would be able to push down the south side of the Kanawha River and procure salt from the Kanawha Salines near Malden.[47] It would seem on initial examination that the Rebel forces were strong enough at this time to meet Lee's expectation. The factors which must be considered in their failure are several in number, but mainly, the poor condition of the troops due to lack of sufficient supplies and medical stores and the fact that so early in the war troops on both sides were "green" and time was needed before they would display the military talents which later crowned their efforts.

General Wise attempted to put the best possible face on the skirmishes and even exaggerated considerably in his official report to General Lee. The Confederates did garner some positive effects from their attacks, in that General Cox became even more cautious and maintained at Gauley Bridge more men than were actually necessary for the protection of that post. This fact proved beneficial to the Confederates when General Rosecrans attacked Floyd at Carnifex Ferry on September 10.

With the exception of a brief skirmish at Cannelton on Thursday, September 5, there was a lull in military activity for several days.[48] Both sides busied themselves in scouting and entrenching their positions. On Sunday, September 8, Cox sent the following dispatch to General Rosecrans, who was then on the march towards Summersville.

No. 17   Gauley Bridge, Sept. 8, 1861
General W. S. Rosecrans:

Nothing from you since your number 28. Wise is now encamped about two miles above Hawks Nest: has three pieces of cannon. His forces occupy about 300 tents, all except officers being the common tent. Some may be in houses, but probably not many. This would make his force not more than 2,500 I think, and will agree with the story of wounded prisoners. We have credible information that they made the attack of the third on New River side with two regiments infantry and one of cavalry. Three regimental colors are flying in their camp. These facts corroborate the prisoners story that their attack was intended to be in force. The scouts report no forces to be found west of Gauley and this side of Peters Creek for ten miles up the streams. Those sent to examine Cross Lanes do not report yet. Nothing new south of the river.

J.D. Cox
Brigadier General, Commanding[49]

    General Floyd had received reliable information concerning the advance of General Rosecrans upon his position at Carnifex Ferry. The animosity between him and Wise was further exasperated when Wise failed to reinforce Floyd in time for Rosecrans's attack on the afternoon of September 10. Floyd's 2,000-man army battled Rosecrans's 7,000-man army until nightfall put an end to the fighting. The Confederates put up a stout resistance against this superior force, and they still held their ground when the battle was over. General Floyd realized that the Yankees would be able to overtake his position the next day. The decision was made to retreat across the Gauley River into Fayette County. That Floyd was able to withdraw his entire force and most of his baggage and equipment without the enemy realizing it was probably the highlight of his military career. Never again would he prove to be so adept at difficult military matters. The Confederates had inflicted comparatively severe injuries to Rosecrans's army, with 27 killed, 103 wounded, and 4 missing. Miraculously, Floyd's forces had none killed, 7 wounded, and 17 captured.[50]

    Both Floyd and Rosecrans claimed victory at Carnifex Ferry, and both had good arguments to support their claims. Despite the indecisiveness as to who won the battle, one important fact remained in favor of General Rosecrans. By forcing the Confederates to abandon Camp Gauley, he reopened the Federal line of communication between Sutton and Gauley Bridge and dealt a severe blow to the Confederate ideal in western Virginia by making it possible for the loyal government of Virginia to meet in Wheeling in October. That meeting paved the way for West Virginia statehood.

    General Floyd's army withdrew along the Sunday road and met General Wise near present day Hico. Floyd was understandably upset that Wise had not reinforced him in time for the battle. His agitation prompted him to write a letter on September 12 to L. P. Walker, the Confederate Secretary of War.[51] Wise's failure to cooperate with Floyd at Carnifex, and Floyd's complaints to Walker and Confederate President Jefferson Davis, eventually resulted in Wise's being relieved of command. Had it not been for Wise's popularity with his Legion and the spectacle that would be presented to the press, he would very likely have been arrested by Floyd.[52]

After conferring with Wise, General Floyd ordered their combined armies to fall back to Big Sewell Mountain, which was done on September 13. Upon reaching Sewell Mountain Floyd's camp was established on the western top, and Wise established Camp Defiance on the eastern top, one mile from Floyd's position. Here, both commanders attempted to supply and refit their commands. Almost daily rainfall had caused the turnpike to become exceedingly difficult to travel, and the lack of sufficient shelter and food increased the spread of disease within their already weakened commands. The militia forces that had been in the vicinity of Fayetteville since late August withdrew on September 14, with Beckley's command returning to their base of operations at Raleigh Courthouse (Beckley).[53]

General Floyd called a council of war on September 16, during which it was determined to fall back to a position on Muddy Creek, near Meadow Bluff. Floyd contended that Muddy Creek was the place to receive the attack of Rosecrans, which he was sure would come. Wise, of course, took the opposite view and was for fighting at Sewell Mountain. Seeing that the decision of the council had gone against his wishes, Wise became exceedingly angry, as described by Col. Henry Heth:

> After the council had adjourned, General Wise got on his horse and rode to his command, where he struck the first detachment, halted, raised himself in his stirrups, and in a stentorian voice called out, 'Who is retreating now? Who is retreating now?' He road slowly on, and seeing another group of his men, he repeated the same to them. Presently, his entire command had assembled and he said, 'Men, who is retreating now? John B. Floyd, God damn him, the bullet hit son of a bitch, he is retreating now.'[54]

Floyd retreated to Muddy Creek, establishing headquarters at Meadow Bluff, Greenbrier County. Although Wise had been ordered to join in the retreat, he sternly refused, keeping his command defiantly at Camp Defiance. In these positions both commanders remained with General Floyd attempting over a period of several days to bring the Wise Legion on to his location. The distance between the two camps was about sixteen miles but the condition of the turnpike was such that it may as well have been sixty miles. Nevertheless, General Wise stubbornly held to his position and refused to budge. The scene of action switched to the Sewell Mountain ranges with the advance of General Cox on September 16, and was described as follows:

> General Orders No. 18   Headquarters Kanawha Brigade
> Sunday Road, September 16, 1861
>
> The troops will move from camp, Sunday Road, at 2:30 p.m. this day, 16th of September, 1861, for Spy Rock. The Eleventh and Twenty Sixth regiments, cavalry, and artillery, will occupy the field on the right of the road beyond Aldersons. Headquarters will be at Aldersons House.
>
> By order:
>
> J. W. Conine,
> Acting Assistant Adjutant General[55]

Once established in his camp at Spy Rock (present-day Lookout), General Cox ordered scouting parties out in the direction of Sewell Mountain and sent a report of his activities to General Rosecrans in Nicholas County.

> Headquarters Kanawha Brigade
> Camp Lookout, September 17, 1861
>
> General W. S. Rosecrans, Camp Scott:
>
> General: We have made no forward movement today, McCook being in expectation of his train. Most of mine has arrived. My advance guard is at the foot of Sewell Mountain, and I expect it to report a reconnaissance to the summit. A scouting party of the enemy was on Sewell last night. Have heard of none nearer.
>
> J. D. Cox
> Brigadier General Commanding[56]

General Cox used the home of seventy-year old George Alderson as his headquarters. This fact did not set well with Alderson, who was a strong Confederate sympathizer, and whose eldest son was a member of the Confederate Congress at Richmond. Alderson's home was a well-known and often used inn and stagecoach stop. The Alderson property was almost entirely dotted with the white tents of the Union Army, and his springs were nearly used up before the campaign in this region was over. In fact, the campaign here would not be over for several weeks and would be continued with the arrival of General Robert E. Lee in Fayette County.

Colonel George Alderson, proprietor of the Alderson Inn, Lookout, WV. The inn closed its doors to the traveling public in 1873.
COURTESY EVA FITZWATER, LOOKOUT, WV

The Alderson Inn and stagecoach stop which stood at Lookout, WV. Built about 1840 it was torn down in 1899. Used as a headquarters and hospital by North and South during the Civil War.
COURTESY EVA FITZWATER, LOOKOUT, WV

ROSTER FOR 10th COMPANY 142nd VA. MILITIA
ORGANIZED IN FAYETTE COUNTY

Joel Abbott
L. R. Abbott
Dickson Arthur
James Arthur
Marshal Arthur
Oliver Arthur
Stephen Arthur
William Arthur
Charles Blake
Claudius Blake
Lewis Blake
William Blake
Thomas G. Bland
George Brafford
William Brower
Henry Burgess
Washington Burgess
Willington Burgess
Samuel Carter
J. S. Cassidy—Captain
R. B. Cassidy
S. H. Cassidy
John A. Dempsey
J. E. Dempsey
A. C. Fellers
Jessie R. Fellers
William C. Fellers
John H. Ford
William H. Franklin
William A. Griffith
William A. Kelly
Spragg Lawrence
Hiram W. Miller
George W. Peebles
Marshal Price
William H. Sandridge
John Saturn
James P. Settle
Joseph Short
Levi Tincher
George Tyree
Calvin S. Warner
Johnathan Weaver
Anderson Wilson
Andrew Wilson
Charles Wilson
Anderson Wood

TOTAL: 47

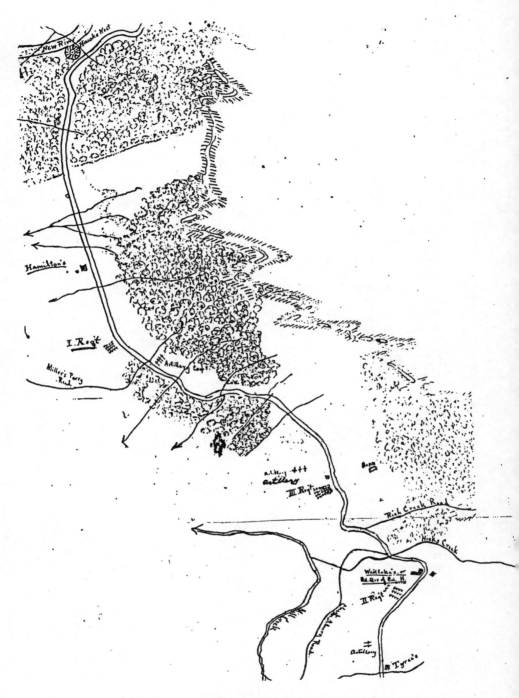

Manuscript map of Hawks Nest and vicinity drawn by a Confederate soldier in 1861. COURTESY VIRGINIA HISTORICAL SOCIETY

■ *Chapter Four*

Robert E. Lee in Fayette County

With the commands of Wise and Floyd separated, both commanders began preparing their camps against attack. General Wise ordered the provisions and baggage wagons withdrawn to safe positions and the camp on all sides strengthened. On September 18 Wise addressed the troops of his legion stating substantially that hitherto he had never retreated but in obedience to superior orders, that here he was determined to make a stand; that his force consisted only of 1,900 infantry and artillery, and the enemy was alleged to be 15,000 strong; that this he did not believe, but that his men must be prepared to fight two or three or several to one, and even if the enemy were in full force stated, the position admitted of successful defense, and he was determined to abide the issue. He warned them that they would probably be attacked front and rear for successive days, and he called on any officer or soldier who felt doubtful of the result, or unwilling to stand by him in this trial, to step forward, promising that they should be marched at once to Meadow Bluff. The speech, delivered successively to the three regiments of infantry and to the artillery, was received with the wildest enthusiasm. Not one solitary individual in the legion failed to respond, and the spirits of the corps were raised and maintained at the highest fighting pitch.[1]

In the retreat from Sewell, General Floyd had ordered Wise to bring up the rear of the column. The fact that Wise failed to do so did not dissuade Floyd from further attempts to unite their forces at Meadow Bluff. In a letter to Wise on September 19, Floyd explained his reason for retreating and expressed his disappointment in Wise:

> I supposed that my order to you of the 16th instant was sufficiently explicit, inasmuch as it therein distinctly stated that I would put at once my column in motion, and that you should hold your command in readiness to bring up the rear, and I have not yet been able to discover how you could bring up the rear of a moving column by remaining stationary after this column had passed. My determination and order to fall back upon the most defensible point between Meadow Bluff and Lewisburg was based upon what I conceived the safety of my command demanded. I felt sure that it was the plan of the enemy to advance upon

General Robert E. Lee as he would have appeared when he arrived in Fayette County. He grew his famous grey beard while at Sewell Mountain. COURTESY SWV

Lewisburg, and in at least two columns, by the turnpike and the Wilderness Road, and to unite their columns at the junction of those roads. In this persuasion I was not mistaken. My scouts on the Wilderness Road have just come in, and report that the enemy are advancing upon that road, which in all probability is true. I felt that this junction could be more certainly prevented and, if effected, could be most successfully met by the combined movement of all the forces under my command.[2]

Wise responded that he had been ordered to be in readiness to move, but that no order to move was ever given. He went on to ask for reinforcements and complained that some of his wagons and supplies had not been returned by Floyd's quartermaster.[3]

The bickering between these two political generals had caused considerable consternation among the civilians of southern West Virginia. President Davis had already received several letters of concern from various prominent citizens in Lewisburg. On September 19, Mason Mathews, Greenbrier County representative to the Virginia Legislature, wrote Davis of his concerns:

I allude to the unfriendly relations existing between the two generals, Floyd and Wise. They are as inimical to each other as men can be, and from their course and actions I am fully satisfied that each of them would be highly gratified to see the other annihilated. I have spent a few days recently in their encampments, and learn that there is great dissatisfaction existing among the officers as well as the privates, and am of the opinion that it would be much better for the service if they were both deposed, and some military general appointed in their stead to take command of both their divisions. This I am sure would be gratifying to the commandants of the different regiments, and would insure success to our cause, at least in this division of our army. It would be just as easy to combine oil and water as to expect a union of action between these gentlemen.[4]

Indeed, both men held stubbornly to their positions and wrote numerous letters of complaint against the other to President Davis and General Lee. Unable or unwilling to rise above their petty jealousies, both contributed more to the loss of western Virginia than the Union army could have done with twice the force then available.

Being convinced that he would eventually fight the Yankees at Meadow Bluff, General Floyd ordered trenches dug along the eastern bank of Meadow River. These trenches were practically worthless for defense as the position was almost entirely surrounded by high hills from which an enemy could easily render the position untenable. His order read:

Special Orders,   Hdqrs. Army of the Kanawha,
No. 97   Camp at Meadow Bluff, September 19, 1861

The Colonels of Twenty Second, Thirty Sixth, Forty Fifth, Fiftieth, and Fifty First Virginia Regiments Volunteers, and of Thirteenth Georgia and Fourtenth North Carolina Regiments, will each upon reciept of this proceed without delay with his entire command and all the entrenching tools in his possession to the bridge across Meadow River, one and one quarter miles west of this point. Col. Henry Heth, Forty Fifth Regiment Virginia Volunteers, will then assign to each command their work. A prompt execution of this order is urged, as the enemy in very large force are advancing upon this point and the works of fortification very incomplete.

By order of Brig. Gen. John B. Floyd:

                                                             Wm. E. Peters,
                         Assistant Adjutant General, Floyds Brigade[5]

While Floyd was constructing useless entrenchments, troops of the Wise Legion were skirmishing with the enemy. From various scouting reports Wise knew that a large force of Yankees had advanced to within six miles of his position. Believing that he might ascertain accurate information as to the strength of the enemy, Wise determined to make an advanced scout, which was carried out during the evening of September 20. He selected five companies of infantry, comprised of the Richmond Blues, commanded by his son, Obadiah Jennings Wise, and the Louisiana Rangers, temporarily commanded by Captain Frank Imboden.

Advancing but a short distance from camp, Wise and his party came under a heavy fire from advance elements of the Union Army. A lively skirmish ensued which was continued on and off all night. Wise was successful in driving back the enemy pickets, and gathering the information he desired as to their strength. The Federals lost one man killed in the fighting; the Confederates had no casualties.

Returning to camp on the afternoon of September 21, Wise was greeted by a dispatch from General Lee, who had just arrived at Meadow Bluff after an unsuccessful campaign against the Yankees at Cheat Mountain, Randolph County. Lee wrote:

> I have just arrived at this camp and regret to find the forces not united. I know nothing of the relative advantages of the points occupied by yourself and General Floyd, but as far as I can judge our united forces are not more than one half of the strength of the enemy. Together they may not be able to withstand his assault. It would be the height of imprudence to submit them separately to his attack. I am told by General Floyd your position is a very strong one. This one I have not examined, but it seems to have the advantage of yours, in commanding the

Wilderness Road and the approach to Lewisburg, which I think is the aim of General Rosecrans. I beg therefore, if not too late, that the troops be united, and that we conquer or die together. You have spoken to me of want of consultation and concert; let that pass until the enemy is driven back, and then, as far as I can, all shall be arranged. I expect this of your magnanimity. Consult that and the interest of our cause, and all will go well. With high respect, your obedient servant,

R. E. Lee, Genl. Comm.[6]

General Wise was offended by the tone and content of Lee's letter, and he responded accordingly.

General R. E. Lee: Fraziers, September 21, 1861 5p.m.

General: I have just returned from feeling the enemy, being out all night and driving in their pickets this morning, and finding their precise position; but, wet, weary, and fatigued as I am, your note reads so much like a rebuke, which I do not think I deserve, that I do not dry or warm my person or lose a moment without replying. I am so desirous to deserve and to have your good opinion and approbation, sir, that you must permit me to be plain in saying that I apprehend you have been told something else besides the fact that my position is a very strong one, and regret that I was not heard before inferences were made of which I cannot consent or correct. In the first place, I consider my force united with that of General Floyd as much as it ever has been, and in a way the most effectual for cooperation. General Floyd has about 3,800, and I about 2,200 men, of all arms, and of these at least 5,500 are efficient. . . . I concur in the imprudence of dividing our forces, but submit, most respectfully, that this is the far stronger position in which to combine, notwithstanding Meadow Bluff is said to command the Wilderness Road. The two roads and the two positions had perhaps better be examined, I respectfully submit, before my judgement is condemned. But sir, I am ready to join General Floyd wherever you command, and you do not say where. I will join him here or at Meadow Bluff. The enemy, while I am writing, has been firing on my pickets, as just reported, from the other side of Big Sewell. I chased him today a half mile beyond Kennys, his reported position day before yesterday, and he is now feigning to advance as I retire to camp. I laugh him to scorn, and do not stop writing, as I know he wishes to retire more now than I do. . . . Where common justice has been done me, I trust I have never failed, and never will, to be generous, and I challenge contradiction of the honest, earnest claim for myself, that no man consults more the interest of the cause, according to his best ability and means, than I do. I am ready to do, suffer, and die for it, and I trust, sir, that I may cite you triumphantly as a chief witness of the truth and justice of that claim whenever and by whomsoever it may be assailed. Any imputation upon my motives or intentions in that respect by my superior would make me, perhaps, no longer a military subordinate of any man who breathes. I am sure you mean to cast no such imputation, whoever else may dare. I trust all will go well, most confidently, in your hands. I am, with the highest respect and esteem, your obedient servant,

Henry A. Wise[7]

Seeing that Wise was strong in his conviction that he had chosen the better of the two positions, General Lee determined to inspect Camp Defiance personally. The following day, the 22nd, Lee rode forward to Sewell Mountain and made a reconnaissance. The position was naturally as strong as Wise had affirmed. It was stronger, in fact, than that of Floyd, sixteen miles to the rear. If the main attack was to be directed along the line of the James River and Kanawha turnpike, across Sewell Mountain, then it was the course of wisdom to bring up Floyd and to fight where Wise stood. After conferring with Wise at length concerning the strength of the position and the preparations that had been made for its defense, Lee left Wise without making a decision or explicitly ordering him to retreat.[8]

The next day, the 23rd, Wise notified Lee at Meadow Bluff that the enemy had occupied the western top of Sewell Mountain with infantry, cavalry, and artillery, all plainly visible from his camp, about one mile distant. He also wrote that considerable skirmishing had taken place, saying: "I am compelled to stand here and fight as long as I can endure and ammunition last. All is at stake with my command, and it shall be sold dearly."[9]

Growing concerned that the enemy might attack Floyd by the Wilderness Road, while demonstrating in front of Wise, Lee instructed Wise to send back his baggage train and prepare for a quick retreat on the first evidence of a move against Floyd.[10]

The situation at Camp Defiance was rapidly growing serious as the skirmishing began to appear more as a general engagement. Colonel Henningson, Lieutenant Colonel Frank Anderson, Captain Imboden, Captain Lewis, and Captain Crane's University Company, were all engaged in the fight with three companies of infantry and two artillery pieces, under Major Gibbs, Captain McComas, and Lieutenant Pairo.[11] The fighting continued from early afternoon until dark. Lieutenant Howell of the 2nd Regiment, and one of Capt. Imboden's Louisiana rangers were killed. Captain Lewis was severley wounded and three privates slightly wounded. The only loss in the artillery was Lieut. Pairo's horse shot under him. Major Lawson of Wise's 2nd Regiment barely escaped injury when an enemy scout put a ball through his coat. The Federals reported no casualties, which seems unlikely considering the nature of the fight. If Confederate reports are to be at least partially believed, the Yankees suffered one killed and several wounded.[12]

The enemy advance, of which Wise had notified General Lee, was, indeed, the movement of several regiments from their position at Spy Rock to a new camp, one mile west of Camp Defiance.[13]

General Orders,   Headquarters Kanawha Brigade,
No. 20   Camp Lookout, September 22, 1861

The troops will move from camp Lookout to-morrow morning, the 23rd of September, 1861, for Big Sewell. The general will be beat from the headuarters at 8 a.m., when every tent will be struck, packed in the wagons, and all fires put out. At 9 a.m. the march will be beat in the infantry and the advance sounded in the cavalry, when each regiment will take its place in column. The order of march will be as the different regiments are encamped. The men will be furnished with cooked rations for two days. Each regiment will leave a sufficient guard to protect the remaining stores in camp.

By order of Brig. Gen. J. D. Cox, commanding:

J. W. Conine,
Acting Assistant Adjutant-General

    During the night of the 23rd General Lee decided to reenforce Wise and return to Camp Defiance himself. Four regiments were selected, including the 13th Georgia, and 14th North Carolina Infantry, and two Virginia infantry regiments, with two pieces of cannon. Orders were given to have tents, baggage, and everything packed in wagons, one day's rations in haversacks, and every man and company ready to start at 5:00 a.m. After a delay of two hours the line of march was taken up at 7:00 a.m.[14]

    General Lee was accompanied by twenty-one-year-old Captain Walter H. Taylor. Captain Taylor had been serving as Lee's only staff officer since the death of Lieut. Colonel John A. Washington, at Elkwater, West Virginia, on September 13.

    A more efficient and dedicated man than Taylor would have been difficult, if not impossible, to find. Though Lee's staff reached a maximum of seven later in the war, the man who "made himself indispensable to Lee's headquarters," who "was first to last the closest," of all staff officers to Lee, was the "extremely efficient" adjutant-general, Walter Taylor.[15]

    As the long column of reenforcements began nearing Camp Defiance it encountered elements of the Wise Legion marching to the hospital at Meadow Bluff. Major Isaac Noyes Smith, of the 22nd Virginia Infantry, recorded the event in his diary: "Marching rapidly, we met men and wagons from Wise's camp at almost every turn. Many of them had great stories to tell of how the pickets were shot etc. the wagons hurrying to the rear, and we met some going rearward very rapidly whom we thought ought to be going forward."

    The stories these men told were not without foundation as there had been several skirmishes in the last few days and another fight was in progress at that very time, as described by Major Smith: "At the top of Sewell or near it (Mrs. Buckingham's old place) halted a moment to rest

Colonel Walter H. Taylor. He was a close friend of General Lee's and became an indispensable member of his staff.
COURTESY NATIONAL ARCHIVES

the men, had hardly sat down when the booming of cannon was heard, the Colonel jumped to his feet, called out 'forward' and we were off again."[16]

Reaching Camp Defiance around 2:00 p.m., the reenforcements were marched off into the woods and shown their place along the top of the ridge. Continual firing among the pickets echoed through the mountain defiles and much anxiety was felt within the camps. With each thunderous blast of artillery the earth shook for several yards around, and many of the new recruits began to wonder for the first time if they would ever see the comforts of home again. Major Smith recorded his feelings in his diary: "It seems so singular that we should be here so near and for so deadly a purpose. I feel so much more like shaking them by the hand, urging them to let us alone, to go home, end this fraticidal war, and whilst they live under their government, to let us live under ours unmolested."[17]

General Lee arrived at Camp Defiance in no good humor, and the situation he found there only made matters worse. Many of the officers were discontented, ignorant of their duties, and bitter toward Floyd and

Scouts on Sewell Mountain, Fayette County. COURTESY SWV

his command. Captain Taylor described the situation. He wrote: "The bitter feeling which had been engendered between the two commanders had imparted itself, in some degree, to the troops, and seriously threatened to impair their efficiency. No little diplomacy was required therefore, to produce harmony and hearty cooperation, where previously had prevailed discord and contention."[18]

General Lee busied himself with examining the terrain and viewing with field glasses the position held by General Cox, one mile distant. It was about this time that Lee was approached by a youthful Lieutenant of a command that had been on Sewell Mountain for several days. This officer calmly asked Lee to tell him who his ordnance officer was and where he could find the ordnance depot. With patience at an end, Lee eyed him sternly: "I think it very strange, Lieutenant, that an officer of this command, which has been here a week, should come to me, who am just arrived, and ask who his ordnance officer is and where to find his ammunition. This is in keeping with everything else I find here, no order, no organization, nobody knows where anything is, no one understands his duty."[19]

Amid this military chaos General Wise strode defiantly, confident his army could whip the Yankees with or without Floyd's help. In one of his brushes in thick woods, Wise ordered an artillerist to open fire. The officer protested that he could not see the enemy and could do no execution. "Damn the execution sir," Wise was reported to have said, "its the noise that we want."[20]

That night Lee bivouacked on the mountainside covered by his overcoat, for his wagon had not come up. It was about this time that he began to grow the beard which would become so familiar to Americans as the classic image of Lee. Coming to Meadow Bluff from Randolph County, he had ridden across country and had no baggage wagon when he reached Floyd; his effects did not arrive until September 26.[21]

Many of the same problems that plagued the Confederates also existed within the camp of General Cox. There was a lack of discipline and a severe need for experienced officers to lead and control the "raw" troops. Though the Federal forces were generally better supplied, there was much waste and inefficiency occasioned by the need of experienced quartermasters. Disease was equally a problem. Hundreds of men were sick with measles, diarrhea, camp fever, and other ailments. Many of the homes and churches between the western crest of Sewell Mountain and Gauley Bridge, were used as hospitals. Food for the men and forage for the horses were rapidly becoming a problem, and the items which were available had already seen a substantial increase in price.

General Rosecrans was concerned that Cox had advanced too near the Rebel position to avoid a fight, and cautioned him to provide against

Captain William Ledbetter of the "Rutherford County Rifles," 1st Rgt. Tennessee Infantry. His regiment served on Sewell Mountain with General Lee.
COURTESY WILLIAM C. LEDBETTER, MURFREESBORO, TENN.

Private Elisha R. Vaughan of Captain Ledbetter's Company, Tennessee Infantry. He was among the brave "Tennessee Foot Soldiers" who served with General Lee on Sewell Mountain.
COURTESY JAMES AND VIRGINIA VAUGHAN, MARTIN, TENN.

everything.²² Cox responded that in his judgment they had come to that position in the nick of time, and were in no serious danger from the Rebels. He also provided Rosecrans a description of their positions: "They hold a ridge which commands the road for nearly half a mile, and have a battery of apparently one rifled four pounder, one smooth sixer, and a mountain howitzer. . . . The intermediate ridge on the left has a road running along its crest, which is said to be barely passable for wagons. The crest they are on is thickly wooded, and I am not yet sure whether it can be reached so as to flank them."²³

On the morning of the 25th, the 13th Georgia Infantry was marched from their position along the ridge, to the bald hill which was on their left flank. This bald hill is today known as Busters Knob. The movement was observed by Cox's men, and he reported it to General Rosecrans: "They still hold the position, and there are some signs of a gun on the bald hill I have mentioned, apparently with a view to protect their left flank. I will keep reconnoitering, but as much as possible without fighting, until you arrive."²⁴

Later that same day General Cox ordered the 11th Ohio Infantry to scout the right side of the Rebel camp. This scout was described by Joshua Horton of the 11th Ohio: "Our regiment was sent on a reconnaissance to the right of the Rebel camp on Big Sewell. Moving over a succession of steep hills, through the rank and tangled undergrowth of a dense forest, we had almost got into the Rebel entrenchments before discovering our position. Companies A and F were deployed in advance, and the men fired on the Rebel pickets whenever they could be seen through the bushes. Louis Brossy and James Mahan, of Company A, getting separated from their comrades, unintentionally walked into the Rebel camp and were made prisoners."²⁵

When the picket firing began, Wise, himself, led a strong force of skirmishers to meet the advancing Federals, and rifle fire was opened at long range. At about 4:00 p.m., while under fire in the sharp skirmishing, an order issued at Richmond on September 20, was handed to General Wise, directing him to turn over all troops under his command to General Floyd, and to report in person at Richmond "with the least delay."²⁶

Explicit as were the terms of the order, Wise debated whether to obey or defy the War Department as he had already defied Floyd. In his hesitation he wrote Lee, asking his opinion. Lee replied promptly and urged Wise to comply with the instructions. Wise drafted a farewell to his men, announcing his recall and affirming that when President Davis instructed he be relieved, he could not have foreseen that the order would be received when the troops were in the face of the enemy in the hourly expectation of a fight.²⁷ Wise immediately packed his baggage and left for Richmond the next morning, reaching that city on the 28th.

Wise departed from Lee's immediate supervision, but he was to serve with Lee again later, and to share the ordeal of Appomattox in precisely the same spirit he displayed on Sewell Mountain.[28]

On the morning of September 26 General Rosecrans arrived in person at the camp of General Cox. He had not been able to bring up his headquarters' train, and for the next two or three days he shared the tent of General Cox, along with General Robert C. Schenck, who reported sometime after Rosecrans. General Cox later described the situation. He wrote: "In my own tent General Rosecrans occupied my camp cot; I had improvised a rough bunk for myself on the other side of the tent, but as General Schenck got in too late for the construction of any better resting place, he was obliged to content himself with a bed made of three or four camp stools set in a row."[29]

A few hours after Rosecrans reached Sewell Mountain a heavy rain began and continued for nearly three days. Disease and supply problems, which were already at near intolerable levels in both camps, became many times worse before the storms ended. The turnpike, which was rarely in good condition, even in dry weather, now seemed to disappear: "The roads were in a dreadful condition; the bottoms appeared to have dropped out."[30] Another witness recorded: "In some places, every trace of the road had been so completely washed away that no one would dream that any had ever been where were then gullies eight or ten or even fifteen feet deep."[31]

A member of the 22nd Virginia Infantry later wrote of the terrible storm: "The rain poured on pitilessly; the poor fellows were shivering in the wet and cold. A more merciless, cheerless rain, and more miserable day could scarcely have been experienced by anyone. The poor, half frozen, half starved men had to stand it. We were all thoroughly drenched, and with difficulty kept the rain from extinguishing our fire."[32]

The temperature dropped sharply and just after dark the rain turned into sleet. Men and horses alike suffered immeasurably and the telegraph lines, which had been installed along the turnpike, were washed out. Amazingly, reenforcements reached Camp Defiance at 10:00 p.m., during the worst of the sleet storm. Peering from their tents, wagons, and whatever shelter they had, the troops strained to identify the approaching column. The reenforcements were found to be the 20th Regiment Mississippi Infantry, who had marched three days from the depot at Jackson River to reach Sewell Mountain. Their arrival was described by J. H. Miller:

> We arrived at Big Sewell Mountain about ten oclock at night in a pelting storm of sleet, without tents, wagons, or any kind of vessel to cook in, the wagon train being left behind coming up the third day after our arrival. We could get plenty of raw rations, beef and flour, but how to cook the bread was a problem we were sometime in solving. We after many

thoughts, struck the keynote by making flour in a dough, drawing it out in a long roll and wrapping it around an iron ram-rod, placing the forks near the fire and laying the ram-rod in the forks, turning the rod as necessary, you soon had a roll of bread nice enough for a lord to dine on; light rolls and broiled beef were our only food for several days and it was much relished in the camps."[33]

The storm subsided a little on the 27th, and Lieut. Colonel James W. Spalding, of the 60th Regiment Virginia Infantry, decided to conduct an advanced scout of the enemy. This scout was totally unauthorized and it was reported that Spalding had been drinking and, in fact, was intoxicated. Ordering out one of his companies, he advanced beyond the Rebel pickets who were posted just 300 yards from the enemy. Carelessly, he rode out in advance of his party and soon found himself face to face with pickets of the 30th Ohio Infantry. General Hugh B. Ewing, of the 30th Ohio, described the event in a letter to his wife: "Col. Spalding rode up the mountain, turned the curve, and suddenly came upon the picket. He reined his horse, drew his pistol and fired, they fired, he missed them, they shot him in the breast. His horse whirled and ran down the mountian, carrying him fifty yards and hurling him to earth in the midst of his escort. They fled carrying off his body. He left at the sport were he fell, a pool of blood, his pocket book and papers, fallen from his breast pocket. All cut by bullets or stained by his hearts blood."[34]

Spalding thus became the third Confederate officer to be killed in Fayette County. His sudden and shocking death was a severe blow to morale among the Confederate troops.

Supplies for Camp Defiance were sent regularly from Floyd's camp at Meadow Bluff. The rains caused additional concern that the men and horses might have to endure supply problems even worse than those they had already experienced in this region. On the 28th Floyd wrote to Lee advising him of his situation: "I understand that the bridges have been carried off in two places between this camp and Sewell, which may delay the transportation of supplies until they can be repaired. The streams are still very high but are rapidly subsiding. They shall be crossed and provisions sent forward as soon as this is possible."[35]

The heavy rains also caused a great deal of damage in the Kanawha Valley. With the Gauley and Kanawha Rivers rising at an alarming rate, there soon occurred a flood which was on a scale never seen before or since. General Cox described the flood in his post war reminiscence. He wrote: "The inundation almost stopped communication, though our quartermasters tried to remedy part of the mischief by forcing light steamers up as near to the Kanawha Falls as possible. But it was very difficult to protect the supplies landed on a muddy bank where were no wharehouses, and no protection but canvas covers stretched over the piles of barrels and boxes of bread and sacks of grain....we managed to keep our men in rations, and were better off than the Confederates."[36]

With the inclement weather there naturally came a lull in skirmishing on Sewell Mountain. It was not uncommon for the pickets to exchange Kanawha salt for Greenbrier beef. They would also discuss the merits of each government's position in the conflict. While the majority of Yankees at Sewell were from Ohio, some had been residents of the Kanawha Valley who enlisted into Ohio regiments.

The severe weather also kept General Lee near his tent, and he spent many hours in conference with his commanders, and entertaining occasional guests. Lee's aide, Walter H. Taylor, later wrote a description of the headquarters on Sewell Mountain: "One solitary tent constituted his headquarters camp; this served for the General and his aide; and when visitors were entertained, as actually occured, the General shared his blankets with his aide, turning over those of the latter to his guest. His dinner service was of tin, tin plates, tin cups, tin bowls, everything of tin, and consequently indestructible; and to the annoyance and disgust of the subordinates, who sighed for porcelain, could not or would not be lost."[37]

On Sunday the 29th, everyone breathed a sigh of relief as the rains had ceased and the sun was once again visible in a clear blue sky. The troops were encouraged to attend church services, which were held in several areas of the camps. Having the Sabbath day to rest and forage, the men welcomed the opportunity to mingle with their friends and write letters to their loved ones. Some of the men struck out on their own in search of food. One such search involved General Lee and his staff, as described by private Leroy W. Cox:

> At one time a bridge washed away, which cut us off from all supplies. For three days we had no rations. One day Bill Catterton, a young fellow by the name of Dodd, and myself, got permission to go out and hunt for something to eat, that is, to buy something if we could find anything to buy. We called it a pirating expidetion. We finally came across one of our headquarters wagons and bartered with the fellow in charge of it for about a quart of flour. He said we could have it for fifty cents. We had no change, nothing but a five dollar Confederate bill among us and he couldnt make change. Bill Catterton, an illiterate mountaineer and a tremendous giant of a man, said, "Here, give me the bill, I'll get it changed." It happened that General Lee was standing nearby and Catterton went straight up to him, addressing him thus, "How do you do, General?" and put out his hand to shake hands with him. I can see that ugly hand now as he offered it to the General. It had one time been mangled by a hog and was terribly disfigured. "How are you, General, and hows your family?" "All well the last time I heard from them," replied the General pleasantly, as he clasped the mountaineers huge misshapen hand. "General, will you please to give me change for a five dollar bill?" The General searched in his pockets and said, "I'm sorry, I havent got it. Maybe one of these other

General Robert E. Lee on his horse "Traveller." Lee first saw this horse while camped in Fayette County. This is a post war view.
COURTESY WASHINGTON & LEE UNIVERSITY

gentlemen could give it to you." He referred to some of his staff officials who were close at hand. They all began going through their pockets and one of them announced that he had it. In the meantime, Dodd and I had stayed back at a respectful distance, viewing with horror this audacity on the part of our rough friend. However, thanks to his very audacity, we were soon in possession of the flour.[38]

On that same day additional reenforcements arrived at Camp Defiance. General William W. Loring brought five regiments from the Cheat Mountain area. These were the 42nd and 48th Virginia Infantry, and the First, Seventh, and Sixteenth Tennessee Infantry.[39] Assisting Loring was General Samuel Read Anderson, a fifty-seven year old veteran of the Mexican War. General Loring was forty-two years old and was also a veteran of the Mexican War, having lost his left arm in the fighting at Mexico City.

The march had not been an easy one for Loring's troops. They had traveled on foot from Big Spring, near present-day Slatyfork, to Sewell Mountain in just one week. The creeks and streams which they had to ford were swollen completely out of their banks and looked more like rivers. Loring, with a portion of the command, arrived first, Anderson and the balance arrived the next day. Captain J. J. Womack, of the 16th Tennessee Infantry, recorded their arrival in his diary. He wrote: "We continued our march today to the top of Sewell Mountain. We had to

Homer Wells, photographer, took this picture of Lee's tree in June 1929. Under this tree General Robert E. Lee had his camp during the Sewell Mountain Campaign in Fayette County in the fall of 1861. Here was where Lee first saw "Traveller," his famous war horse. Tree was a sugar maple and was cut down by the CCC for Daughters of the Confederacy about 1936. Materials from it were made into souvenirs.
COURTESY FAYETTE COUNTY HISTORICAL SOCIETY

cross two small streams today, which, from the rain on Friday, were spread out near a half mile in width, and which the soldiers had to wade, causing us to be late at night when we arrived at the top of the mountain. Here, on the top of this lofty summit we pitched our camps, only one thousand yards from the enemy, who occupied a neighboring hill, separated from us only by a deep ravine."⁴⁰

Another member of the 16th Tennessee Infantry described their arrival on Sewell and the events of Monday morning, September 30:

> Our regiment was ordered to take position on the right wing which we did that night, upon a long high ridge partly fortified. When the morning came we could plainly see Rosecrans army on a mountain immediately in our front about one mile off. We now expected a fight. There they were, with all their army, commanded by their boasted Rosecrans, Cox, and Schenck, and stretched far and wide upon a mountain chain and with their much abused banners of the Union, flying in the air amid a flourish of trumpets and martial music. Our army had a splendid position. It was stretched out for two miles along a high commanding ridge curving to the front, the left being occupied by Floyd's Legion, the right center by Wise's Legion, and the Tennessee troops with a strong force in reserve to the rear of the center.⁴¹

With the reenforcements came additional supply problems, and General Lee wrote to Floyd about his concerns: "The concentration of so large a force will require great energy in the quartermasters and commissary departments. Fifty barrels of flour will be required daily for one item, and provender for the amimals."⁴²

It was about this time that Lee began formulating a plan to attack the Yankees. He wrote General Floyd saying he was afraid the enemy would not attack him, and therefore he wanted to attack them. If Rosecrans did not assume the offensive, the Confederates could do so by several routes to the rear of Rosecrans. Obviously, in a country that presented such formidable natural barriers, it was far better to meet an attack than to deliver one. In addition to Lee's other concerns, the Southern newspapers were clamoring for action. He had already been much maligned by the press for his failed attack on Cheat Mountain three weeks ago.

Over in the Yankee camp Cox and Rosecrans had similar ideas. They also realized it would be preferable to receive an attack than to deliver one, and they contented themselves with scouting and fortifying. Sickness was as much a problem for Rosecrans as for Lee, and, in fact, Rosecrans's force had been reduced from approximately 8,000 to just 5,200 effective.⁴³

By October 1, Rosecrans's commanders were having doubts as to the wisdom of remaining much longer in their present position. Cox wrote of their situation: "Every wagon was put to work hauling supplies and ammunition, even the headquarters baggage wagons and the regimental wagons of the troops, as well those stationed in the rear as those in front. . . . the wagon trains were too small for a condition of things where the teams could hardly haul half loads, and by the first of October we had demonstrated the fact that it was impossible to sustain our army any further from its base unless we could rely upon settled weather and good roads."[44]

A movement to Camp Defiance by General Floyd with 2,000 men was completed on October 1. The force now under Lee's command was approximately 9,000 troops, and as many as twenty cannon.[45] With this strong force the men knew something would have to be done and done soon. Major Isaac Noyes Smith described the emotion of the situation in his diary. He wrote: "I am afraid continually that I shall never see the loved ones at home again. If we attack the enemy there will be terrible slaughter. Why should I again survive when so many are certain to be lost? How little we thought when I left the house on Monday evening that it was our last meeting for so long a time, and very possibly forever. I dread an action more on this account than any other. Floyds troops have joined us here."[46]

The following day, the 2nd, it began raining again and continued all night and through the day on October 3. General Lee had determined to attack Rosecrans as soon as weather permitted, and at 8 p.m., on October 3 an order was issued to be ready for action by dawn of October 4.[47] Lee was confident from information received in scouting reports that Rosecrans was planning an attack. There had been considerable noise from the movement of wagons and artillery the day before, and if Rosecrans did not attack soon, Lee would.

Even before the 8 p.m. order was given, the troops of Lee's command knew something was about to happen. During the afternoon all the wagons were loaded and sent to the rear, and the men were told to prepare for a march with no specific reason given. Understandably, these events caused great anxiety within the camps and the men hoped for a defensive rather than offensive action.[48]

The increased activity in Rosecrans's camp seemed to be a sign of an impending attack. General Lee viewed the movements with his field glasses and felt confident the Federals would soon attack him in his entrenched position. The evening passed and no attack was made. Additional pickets were posted to alarm the camp if the enemy advanced, and the troops bedded down. With the situation as it was, there would be little sleep. Major Smith described the events of October 4: "At quarter past three oclock we were all up, and ready very shortly after to receive the enemy. Every moment we awaited the opening of the enemys guns, and so continued until late in the day. At twelve oclock I thought it more than probable no attack would be made during the day. You have no idea of the excitement such a state of things produces, but I have been so long accustomed to such things that the effect is nothing like it would be to inexperienced persons."[49]

October 5 was a calm clear day and it was spent the same as the preceding day, in hourly expectation of a fight. What the Confederates could not have known was that Rosecrans had already decided to retreat and had made preparations to slip away in the night. General Cox described the movement:

> During the fifth of October our sick and spare baggage were sent back to camp Lookout. Tents were struck at ten oclock in the evening, and the trains sent on their way under escort at eleven. The column moved as soon as the trains were out of the way, except my own brigade, to which was assigned the duty of rear guard. We remained upon the crest of the hill till half past one, the men being formed in line of battle and directed to lie down until time for them to march. . . . It was interesting to observe the effect of this night movement upon the men. Their imagination was excited by the novelty of the situation, and they furnished abundant evidence that the unknown is always, in such cases, the wonderful. The night had cleared off and the stars were out. The Confederate position was eastward from us, and as a bright star rose above the ridge on which the enemy was, we could hear soldiers saying in a low tone to each other, "There goes a fire balloon, it must be a signal, they must have discovered what we are doing." The troops had marched but a mile or two when they overtook part of the wagon train toiling slowly over the steep and slippery hills. Here and there a team would be stalled in the mud, and it looked as if daylight would overtake us before even a tolerably defensible position would be reached. .. . When at last day broke, we were only three or four miles from our camp of the evening before; but we had reached a position which was easily defensible, and where I could halt the brigade and wait for the others to get entirely out of the way. The men broiled their coffee, cooked their breakfast, and rested.[50]

Confederate pickets heard and reported the noise from Rosecrans's camp during the night, but it was dismissed as nothing unusual. Major Isaac Smith recorded the event in his diary. He wrote:

> At one oclock that night I set out alone to visit the pickets, and found that one of the pickets had heard great rumblings of wagons but thought there was nothing unusual as they had heard them nearly every night. The camp was surprised next morning when daylight showed the opposite hill perfectly bare which had been filled with tents and moving men the day before. Colonel Tompkins regretted not having permitted me to report and said he would tell General Lee about it. The General afterwards said I should have advanced the pickets and felt the enemy; just think if I had known that was the way to manage the matter, I should have been the first man to occupy the enemys camp, but live and learn.[51]

General Lee ordered an immediate pursuit of the enemy, which was conducted by a detachment of cavalry and four infantry companies.

The pursuers advanced cautiously until they reached the heights of Sewell Mountain, which had served as Rosecrans's camp. Signs of a precipitate retreat were everywhere in the forms of camp equipage, tents, and provisions, and they captured one prisoner. General Lee and Captain Taylor came up in person and Lee observed the rear guard of the Federals with his field glasses, several miles distant. The cavalry was hurried forward but did not pursue far as obstructions had been placed in the road and their horses were weak. An unknown Confederate writer described the pursuit:

> We pursued the enemy as far as Mr. Tyree's, some eight miles. Mrs. Tyree is quite a favorite with our army. Though both her sons are volunteers, and her husband one of our most reliable scouts, she positively refused to flee from her home on the approach of the enemy. She penned her chickens, hogs, and cattle under her own eye, and armed with nothing but a single gun and a brave spirit, she was determined to stand her ground and protect herself and property. When the enemy approached, they pitched into her chicken coop and garden and she pitched into them. With a pitchfork she ran them out of her house, and returning to the other side of her house, she found several Hessians cutting her cabbage and bearing them off. Snatching up her gun and leveling it at them, they dropped their plunder and retreated in more than double quick time. The circumstances not only occasioned great diversion among the army, but her heroic conduct satisfied them that she could only be robbed by being killed, and they never after troubled anything she had by violence. Mrs. Tyree is a fine lady, of good character, and extraordinary will and nerve. She keeps one of the best hotels in western Virginia, and it is a favorite resort of all travelers.[52]

In their retreat, the Yankees destroyed a vast amount of material so as to expedite their withdrawal. General Henry Benham wrote an account of these events in 1873. He wrote: "The retreat of the rear guard of our command (I presume in fear of a pursuit) partook more of the character of a rout, than of the ordinary retirement of unfought, unbeaten troops. I was told of a large destruction of provisions as they came down the mountain, from this advance camp; that whole wagon loads of beef and pork were tipped over in the deep valleys, down the steep escarped roads; and that men walked ankle deep, in flour from the stoved barrels, for a hundred yards together. Fully $30,000 worth of stores were lost or destroyed."[53]

General Lee drafted a new plan of advance on the very day of Rosecrans's retreat, and delivered the preliminary orders for its execution. The promptness with which he fashioned this plan demonstrated a greater facility than he had thus far exhibited in coordinating subordinates to execute his plans. Lee now wanted to move Floyd to the south side of the Kanawha River and have him advance to a point where he could cut the communications of the Federals on the Gauley. Lee would then attack the enemy on the Gauley, and, with Floyd's help, drive them out of the Kanawha Valley. Before such an advance could be attempted the roads had to be repaired so as to assure the transportation of supplies to the Confederate army. The road from Lewisburg to Sewell was in extremely poor condition and Lee's attention focused on necessary repairs. The portion of the road west of Sewell Creek, in present-day Rainelle, was repaired by troops of Generals Loring and Floyd. The road east of Sewell Creek was repaired by the militia of General Chapman. The troops drained the road thoroughly, opened the waterways, and laid timber over all the soft and muddy portions, to form a flooring.[54]

With the retreat of Rosecrans and the road repaired, General Lee allowed the 1,500 man militia force under General Chapman to be disbanded. Lee told the men they could go home to work their crops, but would be called out again if needed.[55]

Lee's plan to advance Floyd to the Kanawha River did not set well with Colonel Henry Heth, of the 45th Virginia Infantry: "I had a long talk with General Lee and expressed to him my views as to Floyds ability to exercise an independent command. I told him if Floyd was given an independent command it would be merely a question of time when it would be captured; that I did not think the Confederacy could afford to lose three or four thousand men, simply to gratify the ambition of a politician who was as incapable of taking care of his men or fighting them, as a baby.... I have seen too much incompetency in General Floyd to change my views."[56]

Heavy rains had resumed on the sixth, and the following day Lee penned a letter to his wife, who was then at Hot Springs, Virginia:

Sewell's Mountain, October 7, 1861

I received, dear Mary, your letter by Dr. Quintard, with the cotton socks. Both were very acceptable, though the latter I have not yet tried. At the time of their reception the enemy was threatening an attack, which was continued till Saturday night, when under cover of darkness he suddenly withdrew. Your letter of the 2nd, with the yarn socks, four pairs, was handed to me when I was preparing to follow, and I could not at the time attend to either.... I hope, dear Mary, you and daughter, as well as poor little Rob, have derived some benefit from the sanitary baths of the Hot. What does daughter intend to do during the winter? And, indeed, what do you? It is time you were determining. There is no prospect of your returning to Arlington. I think you had better select some comfortable place in the Carolinas or Georgia, and all board together. If Mildred goes to school at Raleigh, why not go there? It is a good opportunity to try a warmer climate for your rheumatism. If I thought our enemies would not make a vigorous move against Richmond, I would recommend to rent a house there. But under these circumstances I would not feel as if you were permanently located if there. I am ignorant where I shall be. In the field somewhere I suspect, so I have little hope of being with you, though I hope to be able to see you. . . . I heard from Fitzhugh the other day. He is well, though his command is greatly reduced by sickness. I wished much to bring him with me; but there is too much cavalry on this line now, and I am dismounting them. I could not, therefore, order more. The weather is almost as bad here as in the mountains I left. There was a drenching rain yesterday, and as I had left my overcoat in camp I was wet from head to foot. It had been raining ever since and is now coming down with a will. But I have my clothes out on the bushes and they will be well washed. The force of the enemy, by a few prisoners captured yesterday and civilians on the road, is put down from 17,000 to 20,000. Some went as high as 22,000. General Floyd thinks 18,000. I do not think it exceeds 9,000 or 10,000, though it exceeds ours. I wish he had attacked us, as I believe he would have been repulsed with great loss. His plan was to attack us at all points at the same time. The rumbling of his wheels, etc., was heard by our pickets, but as that was customary at night in the moving and placing of his cannon, the officer of the day to whom it was reported paid no particular attention to it, supposing it to be preparation for attack in the morning. When day appeared, the bird had flown, and the misfortune was that the reduced condition of our horses for want of provender, exposure to cold rains in these mountains, and want of provisions for the men prevented the vigorous pursuit and following up that was proper. We can only get up provisions from day to day, which paralyses our operations.

I am sorry, as you say, that the movements of the armies cannot keep pace with the expectations of the editors of papers. I know they can regulate matters satisfactorily to themselves on paper. I wish they could do so in the field. No one wishes them more success than I do and would be happy to see them have full swing. I hope something will be done to please them. Give much love to the children and everybody, and believe me

Always yours,
R. E. Lee[57]

The Federal forces had withdrawn to Lookout, or Spy Rock, some twelve miles west of their former encampment at Sewell. Once there it was necessary to use the majority of their wagons for transport of the sick. General Cox reported 426 men sick at Camp Lookout on October 7.[58] It was also necessary to order foraging parties out into the surrounding areas in search of food for the men and horses. One such party was attacked by Rebel cavalry near Lookout on the seventh, with no casualties, though seven soldiers of the 2nd Kentucky Infantry (US) lost their knapsacks.[59]

General Cox determined to fall back to positions at Mountain Cove and Hawks Nest so as to be nearer their supply depot at Gauley Bridge. Accordingly, orders were issued for the movement on October 8 and 9:

General Orders,   Headquarters Camp Lookout,
No. 24   October 8, 1861

The Third Brigade, under command of Brigadier General R. C. Schenck, will move to Mountain Cove. It will move so as to arrive their today, and will there be met by a staff officer to show them their new camping ground. Immediately upon the arrival there, all wagons taken by them not belonging to the brigade will be sent back to this camp. By order of Brig. Gen. J. D. Cox, commanding this camp:

G. M. Bascom
Assistant Adj. General

General Orders,   Headquarters Kanawha Brigade,
No. 26   Camp Lookout, October 9, 1861

The brigade will march tomorrow to the camp near Hawks Nest, and the following order of march will be strictly observed by the commandants of regiments and detachments. The general will be sounded at 6 a.m., at which time tents will be struck, baggage packed, all fires extinguished, and the train take its place in the road in the following order: First, baggage of Cpt. Simmonds artillery; second, headquarters baggage; third, baggage of Second Kentucky Volunteers; fourth, baggage of Eleventh Ohio; fifth, baggage of Twenty Sixth Regiment; sixth, baggage of Pfau's cavalry and McMullins howitzer battery; seventh, baggage of the rear guard.
By order of Brig. Gen. J. D. Cox, commanding:

G. M. Bascom
Assistant Adj. General[60]

On the retreat from Sewell General Rosecrans had taken several regiments and returned to his earlier camp at the Tompkins farm on Gauley Mountain. With the return of Cox's troops there was increased activity between Gauley Bridge and Hawks Nest, as can be seen in a letter written by Ellen W. Tompkins on October 10:

> This place ought to belong to the state, it has such a central, important position. Last night the road was lined with wagons and the noise such we could not sleep, the teamsters quarreling and swearing. The barn now is filled with government property, barrels of pork, sugar, etc. If they retreat they will burn it down no doubt, for when they fell back from Sewell Mountain they burnt tents, and the sugar, coffee, etc., was a foot deep, mixed with mud to prevent the rebels from using it.
> General Winfield Scott does not approve of the march up Sewell Mountain. He did not approve of the rout at Manassas. He well may be angry. It has cost Uncle Sam an immense amount of money and hundreds of sick soldiers, in fact it has demoralized and disheartened them. They return saying that the South has generals, but they have not one capable of leading thirty thousand men. Many talk of going home and all say they must go into winter quarters. Regiments of a thousand cant muster more than five or six hundred active men fit for duty.[61]

Ellen Wilkins Tompkins, wife of Colonel C.Q. Tompkins, 22nd Virginia Infantry. Their home on Gauley Mount was used as a headquarters by Federal troops. COURTESY VALENTINE MUSEUM, RICHMOND, VA

The Tompkins estate on Gauley Mountain. This site is today the Hawks Nest Golf Course. COURTESY SWV

By October 11 General Rosecrans had distributed his brigades in echelon along the turnpike; Schenck's brigade was the most advanced, being ten miles east of Gauley Bridge. McCook's brigade was eight miles out, where the road from Fayetteville, by way of Millers Ferry, came in across New River. General Benham's brigade was six miles from Gauley Bridge, and General Cox, with one regiment, was at the Tompkins farm. The balance of Cox's troops were posted at Gauley Bridge and lower posts where they could protect navigation on the Kanawha.[62]

General Lee's plan to advance Floyd's army into the Kanawha Valley was implemented on October 12:

Orders,   Headquarters Army of the Kanawha
No. -   Camp Sewell, October 11

The army will march tomorrow morning at six oclock in the following order: first, Phillips Legion; second, Second Brigade, under Colonel Tompkins; third, First Brigade, under Col. Heth; fourth, the artillery; fifth, baggage wagons belonging to General Floyds headquarters; sixth, ammunition wagons, ordnance; seventh, hospital wagons; eighth, regimental baggage wagons; ninth, supply train. A detail of one officer and twenty five men will be made from the First Brigade as rear guard. The officer will be instructed to allow neither officer, soldier, nor wagon to fall in rear of the guard. Commanders of regiments will pay particular attention to keeping their men in ranks, and to allow no one to go into houses, or take or destroy property along the road.
By order of Brigadier General Floyd:

H. B. Davidson,
Major and Acting Assistant Adj. General[63]

Floyd's advance from Sewell Mountain was found to be easier said than done. The fact that Rosecrans's army had recently occupied the turnpike made its condition even worse than was expected. There still had not been sufficient dry weather for the road to improve and, indeed, the mud was yet ankle- or knee-deep, depending on location. For several thousand men and several dozen wagons to travel such a turnpike required a herculean effort. Occasionally, a wagon became so thoroughly stuck in the thick mud that it could not be extricated and had to be unloaded and left behind. The physical exertion required to march under such conditions wore the troops down rapidly, and some of the horses were shot after breaking down under the strain of their labors.

Though Floyd's men had grown tired of camp life, not all were glad to leave the security of their entrenched position. The editor of the *Lynchburg Republican*, who was a member of Floyd's staff, described their departure:

> We felt sad in leaving our entrenched position on Sewell, where we expected our little army to have covered itself again with the laurels of a glorious victory. But it may be that our work has not been in vain, as the

continually changing tide of war may yet float us back to them, and make them of great importance in our defense. They cover a space of about four miles, and though temporary in their character, were never the less quite formidable. The earthen portion of them on Burwells Mount will probably remain for a century, and be the object of curious interest to our childrens children. It was from this point, they may exclaim, that the tide of Northern aggression was turned back in the war of our independence in 1861.

The same writer went on to describe the movement of an army:

> Did you ever see an army in motion? It is a most interesting and imposing sight, though the most tedious mode of traveling in the world. A regiment moves off in double die, led by its Colonel, and is followed by a long train of lumberlog wagons. Then comes another regiment, followed by its baggage train; then comes the artillery, then another regiment and its train, and so on, alternately, the line moves for miles out of your sight. If a single wagon stalls the whole rear train has to stop until the vehicle is dragged out of the mud, for in many places the road is so narrow that not even a horse, and sometimes not a footman, can pass a single wagon. The consequence is we move about ten miles a day, and when night comes both men and horses are well broken down with the excessive labors of the day. We then have our horses to feed, our beeves to butcher, our tents to pitch, and our suppers to cook. When the road and the weather are good we make 15 or 20 miles a day with less fatigue and trouble.[64]

With the departure of Floyd's army on October 12, General Lee ordered the Sixteenth Regiment, Tennessee Infantry, to occupy the eastern base of Sewell Mountain, some three or four miles from the heights of Lee's main camp. This movement to the base of the mountain was welcomed by the Tennessee troops as they had suffered additional hardships by the frequent blowing of cold winds across the crest of Sewell Mountain. Once in their new camp some of the men slipped away in search of food, as described by Carroll Clark, of Company I, 16th Tennessee:

> That night, a few of us decided to slip through the guard line, go out and get some potato pumpkins, which were numerous in that section, and when baked, were as good as the old yellow yam sweet potato. We cut sticks about four feet long and sharped each end of the stick. We found plenty of pumpkins in a field, stuck a pumpkin on each end of the stick, then shouldered the sticks and started for camp. In going out of the field; I was the last one to cross the fence, and in a few steps further, someone in the bushes near the path yelled out, "Oh, yes damn you, weve got you." The boys in front of me moved on in a hurry and I was in a loap when off fell one of my pumpkins, and of course down went the other, but I never halted, but pulled for camp. I supposed the fellow who scared me so bad, was the owner of the field of pumpkins, but next day I found out that it was some of our own men, out foraging, and decided to have some fun, and they had it.[65]

Area occupied as the main Sewell Mountain headquarters camp of General Robert E. Lee in 1861. The rise in the foreground is a trench dug by Lee's army. COURTESY DAVID MILES, CHARMCO, WV

The following day, the 13th, the 13th Regiment, Georgia Infantry, left Sewell Mountain for Beckley, then known as Raleigh Court House. They reached that point on Thursday the 24th. Also on October 13 as Floyd's troops advanced toward the Kanawha Valley, a small band of Confederate rangers skirmished with Federal cavalry at Fayetteville:

> Fayette C. H. Va. Oct 14th 1861
> Genl Lee
>   Commanding Va Troops
>
> Dear Sir, I left Gwinn Springs by order of Col. Clarkston with a squad of ten men with instructions to proceed to Raleigh thence to Fayette and thence to Cotton Hill for the purpose of learning the position of the enemy. Nothing of interest occured until Tuesday evening when within about 4 miles of Fayetteville I met two mounted men who informed me that the enemy was marching on Fayetteville with a strong force. Learning from them that I could reach Fayette before the enemy, I passed on until I came to the road leading down Laurel and finding that Col. Woods with some 15 of the citizens had passed down the Laurel road in

search of the enemy. I moved on, overtook, and passed him, and scouted the road as far as Cassidys Mill, 3 miles from Fayetteville, and finding nothing of the enemy returned to Fayetteville, arrived at ten oclock. The next morning with Col. Woods, J J Stiles, J W Phillips, Peyton Morton and others, started for Warners to get a sight of the enemy. There is a road leading from Warners to the river opposite the enemys encampment, the road from Warners to the top of the mountain is three fourths of a mile, and from there to the enemy about the same distance. Col. Woods thinks it would not be difficult to get some artillery up to this point. There was 4 camps in reach of artillery from that point on Saturday.

The eastern camp at Vaughns on Sunday morning had been moved, the next encampment is at Eli Woods the next at Hamiltons. Mr. Montgomery arrived here this morning from Cannelton and reports the enemy 9,000 thousand 500 strong, the information is believed to be reliable among the citizens, he will accompany this and explain all. On Sunday about 3 oclock our picket on the turnpike about 3 miles from Fayette C. H. was driven in by about one hundred and eighty men, we sent our horses one mile from town and took a position in a point of woods just below the C. H., there with our little band numbering 21 men we awaited the invaders. As they came within range of our rifles we fired at them, only six shots when they turned and fled for their lives. Our party was so small that we did not think it advisable to expose ourselves, they having already declared their intention to burn and destroy all property within their reach. Some of them were severely wounded one it is thought dead. I have sent a scout down the road 3 miles this morning, I have reported twice to Col. Clarkston at Raleigh but have received no answer.

Yours Respectfully
S.B. Hawley 3rd Sgt.
Caskies Rangers
Wise Legion[66]

On October 15 General Lee sent the 14th Regiment North Carolina Infantry back to Meadow Bluff. That regiment had entered Virginia seven hundred and fifty strong just three months previously, and was now reduced by sickness to just 277 privates fit for duty.[67] Also on the 15th, Lee wrote Floyd, saying that a spy of his had crossed into Rosecrans's camp at Gauley, using a pass issued him by General Rosecrans through his provost marshal, Major Joseph Darr Jr. The spy supplied Lee with detailed, though not precise, information as to the strength of the Federals. Lee also wrote about the lack of provisions on Sewell and the condition of the Wise Legion: "We barely get bread from day to day. No forage. I should have advanced toward Gauley, had it been possible to take the road, with a view of harassing the enemy and damaging his retreat. I sent the quartermaster and commissary on the road to see what could be procured and they report literally nothing. The hospitals in rear are full to overflowing. The men of the Wise Legion are suffering much for want of clothing. The horses are without provender."[68]

The day after Floyd received General Lee's letter he wrote to the Confederate Secretary of War, advising him as to the condition of his command and his plans for winter quarters. General Floyd was more adept at making plans than executing them, and he again displayed his tendency to "bite off more than he could chew," when he suggested that his command be allowed to winter at Logan Court House. This plan was unrealistic, as he would have had to transport supplies 130 miles from the railroad depot on Jackson River: "We remained (on Sewell) eleven days, and those days cost us more men, sick and dead, than the battle of Manassas Plains. Provisions were hauled up the mountain 16 miles from Meadow Bluff over the worst road in Virginia, and we were exposed to tempest of wind and rain; for the conformation of the ground is such that there are always storms on Sewell Mountain. Finally the enemy retired beyond Gauley.... I still adhere to my original purpose of wintering near Logan Court House...."[69]

With winter approaching, General Lee realized that he could not risk the health of his troops any further in this region. He did not share Floyd's desire to winter in Logan County, and in fact began focusing his attention eastward. The large numbers of sick troops who had been at Camp Defiance were almost entirely gone by October 18. Lee had sent them to the hospitals at Lewisburg, White Sulphur Springs, and Blue Sulphur Springs. One such ambulance journey carried among its passengers two members of the 42nd Regiment, Virginia infantry. Captain Samuel J. Mullins, age twenty-nine, and Private John S. Penn, age nineteen. Both men had been ill with "Camp fever" for several days, and though they were treated with large quantities of quinine, their condition worsened. Mullins recorded their departure from Camp Defiance in his diary: "Friday, October 18, 1861, I got my furlough fixed by Dr. Morris's aid. I left in an ambulance with John Penn. We came six miles. Saturday, 19th, We traveled on towards Lewisburg. Sunday, 20th, We got to Lewisburg today. I left John Penn there where he soon died."[70]

The story of John Penn and his brothers is an interesting one in that they were a typical "Confederate family." John Penn and his three brothers had all enlisted into Company H, 42nd Virginia Infantry - all but one brother enlisting together, May 22, at Spoon Creek, Patrick County, Va. There were Joseph, age twenty-eight, Thomas, twenty-two, William, twenty-one, and John, nineteen. William had been the first of the brothers to succumb to disease. He was discharged due to disability on September 18, at Big Spring, Pocahontas County. When John died at Lewisburg in October, Thomas was allowed to escort his brother's remains home to Patrick County. Joseph was furloughed due to sickness in December, 1861, and upon his return in March 1862, he was again

The Penn brothers of the 42nd Virginia Infantry. Seated—left to right: Joseph G. Penn and Thomas G. Penn. Standing—left to right: John S. Penn and William A. Penn.
COURTESY MRS. FRED WOODSON, MARTINSVILLE, VA

furloughed and never returned. Thomas remained at home until April 1, 1862, and was later promoted to first sergeant. He provided a substitute in July 1862, and was discharged. Records do not show whether or not Joseph later died from disease acquired in the army, or if William improved after his release the previous fall. Needless to say, the sacrifices which the Penn brothers made while defending their beliefs exemplified the cycle of pain endured by thousands of families during the cataclysm of Civil War.[71]

On October 20, Lee wrote to Floyd advising him that he could no longer retain General Loring's command in this region. General H. R. Jackson and General Donelson had sent Lee repeated requests for the return of Loring to eastern Virginia, saying that their line of operations was threatened by the enemy and reenforcements were needed. Lee explained that since he had not heard from Floyd as to when he expected to make a movement down the Kanawha Valley, it was improper to maintain such a large force in an area of inactivity. The Yankees had reopened activity in the direction of the Staunton and Virginia Central Railroad, and priority shifted to that theater of operations.[72]

The following day, October 21, Lee and Loring departed Sewell Mountain, never to return. Loring proceeded to the Greenbrier River and Lee encamped at Meadow Bluff. At this same time General Floyd's forces reached Fayetteville after a march of ten days. The Confederate troops found Fayetteville in a frightful condition, following a raid by the Yankees, as described by an unknown Confederate writer: "This whole town is devastated. The Yankees were here on the 19th, and burnt one store and set fire to several buildings, the Court House among them, all of which were extinguished by the inhabitants. They entered private residences and plundered and broke up every article of furniture, carrying off everything that could benefit themselves. I deeply sympathize with the people of this place. The few who are left are very kind to us, doing everything in their power to add to our comfort."[73]

Floyd's troops did not remain in Fayetteville long, but pressed onward in search of the enemy, whom they soon found. An unknown Confederate writer described their advance and subsequent skirmish. He wrote:

> After ten days hard marching over almost impassable roads, building boats, making roads, and constructing bridges, the command reached Fayetteville the 21st, about noon, and without halting marched to the junction of the Millers Ferry road with the Raleigh turnpike, some three miles west of Fayette Court House. The enemy had pickets stationed near the junction of these roads. The advance was halted here, and a part of Colonel Phillips cavalry dismounted, some seventy in number, as skirmishers, commanded by Col. Phillips in person. General Floyd, with his staff, proceeded down the Millers Ferry road about one mile, the pickets of the enemy retreating before our skirmishers without even firing a gun. Here the General was stopped by a blockade of the narrow pass in the river bluffs, made by the enemy. The militia and those who followed to the ferry, some three quarters of a mile further, had to clambor over timber and fallen trees across the road on foot. The road is very narrow and down a deep defile between the bluffs to the river bank. The pickets of the enemy crossed the river in double quick, and commenced a brisk fire from the opposite heights, at two p.m., which was returned by our skirmishers and kept up until sundown. The enemy brought to the aid of their infantry two pieces of artillery, which were used in firing without effect upon our men until the fight closed at night. Our loss was one killed and three wounded slightly; the damage to the enemy was considerble judging from the screams of their men, occasioned by the shots of our skirmishers, who used their Sharps carbines with telling effect upon their artillerists.[74]

After the fighting General Floyd established his camp at the Dickerson farm, near the junction of the Raleigh and Millers Ferry roads. In this position he controlled movement from across New River towards Fayetteville and other points south. Within a few days Floyd's brigade quartermaster, Colonel Isaac B. Dunn, established a daily line of express from Cotton Hill to the depot at Dublin, Virginia, a distance of 115 miles. The express made an average of seven miles per hour, at stages of five miles each. This express service was conducted by unarmed cavalry.

On October 23, General Floyd wrote to the Confederate Secretary of War, from Camp Dickerson:

> ... Their present position is admirably selected.... In this position, the fork of the Gauley and New Rivers, they command the Kanawha River, by which steamboats laden with supplies come within six miles of their headquarters, as I witnessed today with my own eyes. They command, also, the roads to Clarksburg and the northwest, which they have put in perfect order by employing on them the labor of all their prisoners and all the secessionist in the country which they have overrun. In this position, also, they are always ready to strike Lewisburg whenever the Confederate force at Sewell Mountain and Meadow Bluff is removed. To keep their position is clearly their most important object and purpose in Western Virginia. To dislodge them is equally important to us.[75]

Floyd's advance to Camp Dickerson caused General Rosecrans to hasten forward the work of clothing and paying his men, recruiting his teams and bringing back to the ranks the soldiers whom exposure had sent to the hospital. Rosecrans had received reliable information earlier in the month that Lee planned to move against him while General Floyd advanced from the south side of the Kanawha River. Believing that Floyd's advance was a sign of this attack, Rosecrans remained quiet and expectant for several days, awaiting the development of events.[76]

On October 27 Floyd wrote another letter to the Secretary of War:

> ... I am now preparing batteries on the mountian side which will command the road along the river to the enemys camp, by which they receive supplies after they leave the steamboat. I hope to open fire tomorrow morning, and think that they will cause such serious inconvenience and injury, that the enemy will perhaps cross and give me battle under the conditions which I demand for success. But if the enemy will not do so, his force is so powerful, and mine so small, that I shall be unable to do anything with him unless the Department can prevail on General Lee, to make a movement against his front. My march to this point is only part of a larger plan....I have done my part of this work, but I have not heard of General Lee's movements, and unless he should make them speedily, I fear that this campaign must end without any decisive result, and that all the force lately assembled around Sewell Mountain will be of no profit to the war.[77]

Floyd's claim that he had not heard of Lee's movements was false.

The Dickerson House, used by Blue and Grey as a headquarters and a hospital.
AUTHOR'S COLLECTION

Lee informed him on October 20 that he was sending Loring's army away and that he would send the Wise Legion, who were in extremely poor condition, to Meadow Bluff. It appears that Floyd either forgot about Lee's correspondence or intentionally attempted to deceive the Secretary of War. Perhaps Floyd secretly felt that the opportunity of Confederate success in this region was past, and wanted to shield himself from blame.

General Lee was still at Meadow Bluff, attempting to supply the troops remaining, and caring as best he could for the sick. On Monday, October 28, Lee received a letter from Floyd: "All my forces having arrived, I am now ready for active operations. I have possession of Cotton Hill, and stop all the ferries on New River. I have cannon on the heights, commanding Montgomerys Ferry, and I can cut the road up the Kanawha, by which alone the enemy receives his supplies....If you will now make a decided movement in advance with the army at Sewell Mountains, it is nearly certain that we will capture the whole of the Northern army, or drive it entirely from the valley."[78]

To this somewhat surprising letter from Floyd, General Lee responded that he advised Floyd on the 20th of Loring's return to the Huntersville line, that all the sick had been sent on to the various hospitals, that he would be leaving Meadow Bluff that day (the 29th) and visit the hospitals at Lewisburg and White Sulphur, and then proceed to Richmond. He explained further that Colonel J. Lucius Davis, was in command of the troops at Meadow Bluff.[79]

With this last dispatch to Floyd, Robert E. Lee departed the Fayette-Greenbrier County area. As was the case after the failure of his Cheat Mountain campaign in September, Lee would be much abused by the press for his lack of a decisive victory in western Virginia. Colonel Walter H. Taylor described the situation well:

> We had now reached the latter days of October; the lateness of the season and the condition of the roads precluded the idea of earnest aggressive operations, and the campaign in Western Virginia was virtually concluded. Judged from its results, it must be confessed that this series of operations was a failure. At its conclusion a large portion of the state was in possession of the Federals, including the rich valleys of the Ohio and Kanawha Rivers, and so remained until the close of the war. For this, however, General Lee cannot be reasonably held accountable. Disaster had befallen the Confederate arms, and the worst had been accomplished, before he reached the theatre of operations; the Alleghanies then constituted the dividing line between the hostile forces, and in this network of mountains, sterile and rendered absolutely impracticable by a prolonged season of rain, nature had provided an insurmountable barrier to operations in the transmontane country.[80]

Lee's departure did indeed signal the loss of western Virginia to the Confederacy. Already, on October 24, a majority had voted to establish a separate state, and West Virginia was forever lost to the Confederacy and the Old Dominion. General Lee reached Richmond on the afternoon of October 31, accompanied by his aide, Walter Taylor. At a later date Lee was asked by General William E. Starke, why he did not attack Rosecrans on Sewell Mountain. Lee responded that a battle would have been without substantial result, that the Confederates were seventy miles from their rail base, that the roads were almost impassable, that it would have been difficult to procure two days' food, and that if he had attacked and beaten Rosecrans, he would have been compelled to retire because he could not provision his army. "But," said Starke, "your reputation was suffering, the press was denouncing you, your own state was losing confidence in you, and the army needed a victory to add to its enthusiasm." Lee only smiled sadly: "I could not afford to sacrifice the lives of five or six hundred of my people to silence public clamor," he said. And there he left it.[81]

With these events, the story of Robert E. Lee in Fayette County passed into history. The star of General Lee had not yet risen when he turned his horse eastward on October 30. No one could have foreseen the sanguinary conflicts which would rock the divided nation to its very foundation before climaxing in the "Stillness at Appomattox," April 9, 1865.

■ Chapter Five

The Siege of Gauley Bridge

On October 31 a small party of Federal scouts attacked one of Floyd's camps, which was situated at the Huddleston farm along the road to Fayetteville, about one and one-half miles above the Kanawha River. Major Isaac Noyes Smith, of the 22nd Virginia Infantry, described the attack. He wrote:

> ... we were all setting quietly in Huddlestons house, the men lounging about outside and in without their guns....Suddenly the rattling of musketry was heard all around us, and the greatest uproar and confusion prevailed. Men were crowding in the house, and others rushing out all confused and terrified. I was up stairs, and seizing my sword and pistol, rushed out, to make my way down; the narrow staircase was a perfect jam, men rushing up with and after their guns, and others seeking to get out. I found it impossible to get out. An effort to do so would have impaled me upon the bayonets, which were bristling above the heads of the men. Finally forced my way through the mass and rushed onto the road. To my surprise and mortification, the men were flying in every direction. I had no idea where the enemy was, supposed they occupied the hills on both sides of the hollow, and had surrounded us—couldnt see the enemy though. They were firing continually, and some of our men were firing back. . . . Just then for the first time, I saw the enemy though I had been exposed to their fire for some time, and immediately jumped upon the stone wall and called upon the men to follow me, and charge up the hill upon them. . . : the boys came up following me, and the enemy who had begun to run were soon out of sight. They left one fellow dead in the field shot in the center of the forehead, and one mortally wounded through the body. We had one man shot slightly through the foot and another scratched in the hand. The house had a number of bullets through it. I felt great pity for the poor wounded man, he was in great agony—about 20 and fine looking, fair skin, dark hair, and intelligent face. I talked to him kindly—he said he would soon die.[1]

By November 1861 the Union troops stationed at Gauley Bridge had been lulled into a false sense of security. The lack of active campaigning and the natural beauty of the location were such that many of the troops could scarcely believe they were at war. The sheer mountain cliffs, majestic peaks, and deep ravines were like a magnet to the young Ohio soldiers. It was not uncommon for many of them to spend their spare time exploring the mountains and searching for the best vantage point from which to view the wondrous panorama afforded them.

101

Isaac Noyes Smith,
22nd Virginia Infantry.
COURTESY ISAAC N. SMITH
JR., CHARLESTON, WV

The florescent colors of the fall season had been of a magnitude never before witnessed by many of these boy soldiers. Even the seasoned veterans, who had traveled the Kanawha Valley in the days before the war, viewed with awe the beautiful display of fall colors. All these factors contributed to the great shock of the events they were soon to experience.

The early morning hours of November 1, 1861, found everything peaceful and proceeding as usual within the Yankee camp at Gauley Bridge. The fog was not yet off the river, when reveille was sounded. The valley reverberated with the sounds of officers barking commands to the excited troops. Men scurried to and fro, fires had to be made, roll call had to be taken. The troops were especially quick in their preparations for the day's activities because it was payday.[2]

The soldiers posted up and down the river from Gauley Mountain to Cannelton had known for several days that the Confederate troops on the south side of the river at Cotton Hill had been unusually active. On October 30, men of the 2nd Kentucky Infantry (US) had observed large numbers of Rebels working in the mountains, apparently putting up breastworks. At night they could see large fires here and there, and lights were constantly moving around among the trees.³

Little did they know, General Floyd was busily preparing a surprise for them. He had with him on Cotton Hill, a force of approximately 4,000 soldiers, consisting of: First Brigade Virginia Volunteers—Colonel Heth; 22nd Regiment Virginia Infantry—Colonel C. Q. Tompkins; Third Brigade Virginia Volunteers—Colonel Dan R. Russell; Phillips Legion Georgia Cavalry—Colonel William Phillips; 36th Regiment Virginia Infantry—Colonel John McCausland; 51st Regiment Virginia Infantry—Colonel G. C. Wharton; 13th Regiment Georgia Infantry—Colonel W. Ector & Lieut. Col. M. Douglas; 4th Louisiana Infantry—Major George C. Waddell; 20th Regiment Mississippi Infantry—Captain H. Canty; 45th Regiment Commissary—Captain Samuel H. Henry; and Adjutant General—Captain R. H. Finney.⁴

Map of the vicinity of Gauley Bridge, Virginia, which appeared in the New York Times, November 23, 1861. AUTHOR'S COLLECTION

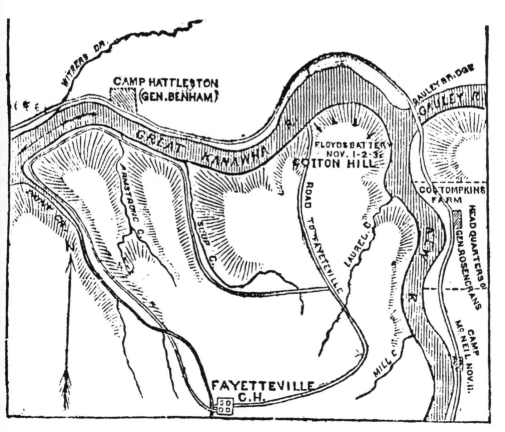

Colonel Henry Heth, commander of the 45th Virginia Infantry.
COURTESY NATIONAL ARCHIVES

Along about 9:00 a.m. soldiers of the Eleventh Ohio Volunteer Infantry, were in line at the paymaster's tent. The men were joking, and much horse-play was evident up and down the line. Suddenly a gun was heard, fired from the south side of New River, and a shell struck in the camp of the regiment.[5]

General Floyd's Confederates had succeeded in posting a rifled cannon on the heights of Cotton Hill, opposite Gauley Bridge, and below Kanawha Falls, opposite Montgomerys ferry. They were now firing their artillery directly into the Yankee camp.[6]

The tremendous surprise of the first shell had not yet passed when the second shell struck. Suddenly the entire scene was transformed into one of total chaos. The telegraph operator, who was housed in a small building near the camp, rapidly sent a message to the headquarters at Gauley Mount. He then abandoned his little post without waiting for a reply. The paymaster promptly moved himself and his money to a safer place. Men were running in every direction, while their officers were vainly attempting to restore order and seek cover at the same time. The shells came fast, like iron rain. With each impact, dirt, twigs, bark and rocks, were scattered in every direction.

Chaplain Du Bois, of the 11th Ohio Infantry, raced from his tent to a nearby tree stump for protection. Old Mrs. White, who assisted the camp cook with the cuisine department, ran with all the speed she could muster, towards a ravine north of camp. Unfortunately, she became entangled in the folds of a quilt she had tossed around her shoulders, and tumbled to the ground in a somewhat demoralized condition. The shells continued to fall within the camp and suddenly one shell came dangerously close to the ferry. It was then that a messenger was sent in a rush to Rosecrans's camp on Gauley Mountain.

Captain Lane, who normally commanded the ferry, was quickly found and asked what to do. He replied that he had no authority in the matter, that the messenger should have gone to Colonel Devilliers. Racing away, the messenger soon found the Colonel, who did nothing but send him back to Captain Lane, saying he should do what he thought best. While all this shirking of responsibility was going on, the soldiers at the bridge abandoned operation of the ferry and moved their wagons out of harm's way.[7]

The Confederates also succeeded in posting their riflemen along the riverbank, east and west of Gauley Bridge. Their firing was quick and accurate, the distance less than 150 yards. After nearly one hour of continuous abuse by the Rebel artillery and sharpshooters, two hundred men of the 2nd Kentucky, worked their way behind trees and rocks along the road, and were busily returning bullet for bullet. The engagement from one side of the river to the other quickly became so general that each side of the river seemed one vast sheet of flame.[8] When the attack began reenforcements were sent to Gauley Bridge from Camp Tompkins, two miles east. Movement along this stretch of turnpike was hazardous, as the Rebel sharpshooters lined the south side of the New and Kanawha Rivers. One of the Union officers described the battle in a letter to his father:

> We as a body moved on towards Gauley, which we reached at about six p.m., having experienced a running fight, in which bullets rained almost, and the loud reports of cannonading sounded terrible. This was my first fight and I really enjoyed the sight and the din of battle. We were obliged to march on an open road, where the secesh had full sweep at us, but strange to say, we had but five men wounded; one of whom is a Lt. in our regiment and another fine looking fellow was shot down by my side in the morning.... While I was lying on my belly watching for a chance to get a shot, a ball struck a rock by the side of my foot and I am free to confess, my sensations were peculiar. After being under fire for sometime, every man of us became so accustomed to the noise, that we seemed to take little thought of the present, but were bent upon giving them thunder, as we certinly did.[9]

Back at the ferry, Captain Lane had got together three volunteers to help him release the ferry and haul it out of range. Working as rapidly as possible, and under direct fire of the Rebel battery, Lane's small crew succeeded in pulling the ferry up river and hiding it behind a projecting point of rocks. No one was injured in the move, although two shells struck close and one was wild. The Confederates were very much amused by all this, but their amusement was almost over.[10]

General Rosecrans ordered a battery of three mountain howitzers to take position and return fire. The first shot was a round six-pounder, which struck the ground close to Jim Grey, of the Virginia Partisan Rangers, throwing dirt in his face. The next shot was a twelve-pound ball, which exploded among the Confederates without hurting any of them. And so commenced one of the most exciting artillery duels ever to be seen in West Virginia.[11]

The shells flew fast and thick, shot answered shot, and soon the entire scene was enveloped in a dense curtain of smoke. So loud was the combined roar from the artillery and musket fire, that it was reported one could hear the battle as far west as Cannelton and as far east as Ansted, a distance of over eight miles each way. The fighting raged almost continuously, the artillery on both sides of the river belching fire and destruction.[12]

Little could be done by the men not manning the artillery or posted along the river. For them the battle was a scene of immense grandeur. Colonel Rutherford B. Hayes, a witness to the battle, described the scene in a letter to his wife the next day, November 2. "I wish you could see such a battle. No danger and yet enough sense of peril to make all engaged very enthusiastic. The echoes of the cannon and bursting shells through the mountain defiles were wonderful."[13]

For many of the troops at Gauley Bridge, this was their baptism of fire. Others had been in minor skirmishes but had never witnessed a battle of such ferocity. It was only the distance between combatants and the inexperience of the troops that prevented loss of life. Finally, after fighting from 9 a.m. to 7 p.m., nightfall put an end to the contest. So weary were the Yankee gunners that they dropped down alongside their cannons. The riflemen who had been fighting along the river's edge were also exhausted. Across the river, the weary Confederates settled down for the night and congratulated themselves on their successful surprise of the enemy. The Rebels also slept beside their cannon, and tried to make themselves as comfortable as possible. Before long a cold rain began and continued off and on all night.[14]

General Rosecrans, being somewhat vexed at his predicament, was busily making plans for the next day's fighting. He ordered General Cox to have the ferry brought back down and operated at night. He also instructed Cox to take whatever measures he thought best to protect their ammunition supply, which had been exposed to the Rebel artillery fire, as it was stored out in the open. Arriving at Gauley Bridge just after dark, Cox ordered the ordnance officer to load their ammunition into wagons and haul it up Scrabble Creek. He also made sure the ferry was set back in operation. The soldiers working the ferry performed double duty and were able to complete a full day's work before dawn.[15]

At midnight General Rosecrans summoned Colonel Sedgewick, of the 2nd Kentucky Infantry, and told him to select about three hundred of his best sharpshooters, and post them along the river road before daylight. Careful not to alarm the Rebels, the riflemen moved stealthily out of camp. The men were told not to talk nor light any torches, for fear the Confederates would discover their activity. The night was exceedingly dark due to the rain clouds and mist along the rivers, and these conditions further complicated an already difficult assignment. After several hours word was passed that the movement was complete, and when dawn broke the men were secreted behind rocks, trees, logs, anything that would protect them.[16]

General Floyd was also active at dawn, sending orders for a wing of breastworks to be built on each side of the cannon. The Rebels had been busy with this work for several hours and were almost finished, when at 8:00 a.m., the Yankees opened fire. The Confederates jumped to the back side of the ridge and sought protection as best they could. Suddenly, one of the shells burst among them, striking the hat of James Gillespie, and passing between two other men.[17] Quickly regaining their composure, the Rebels began firing at some wagons they could see in the vicinity of

Scrabble Creek. These were the wagons transferring ammunition to a safe haven up Scrabble Creek, a job that was to have been completed before dawn. General Cox described the early morning events of November 2:

> The ordnance stores had not been loaded upon the waiting wagons until nearly daylight, and soon after turning out of the Kanawha road into that of the Gauley, the mules of a team near the head of the train balked, and the whole had been brought to a standstill. The line of the road was enfiladed by the enemys cannon, the morning fog in the valley was beginning to lift under the influence of the rising sun, and as soon as the situation was discovered we might reckon upon receiving the fire of the Cotton Hill battery. The wagon drivers realized the danger of handling an ammunition train under such circumstances and began to be nervous, whilst the onlookers not connected with the duty made haste to get of harms way. . . . Without warning, a ball struck in the road near us and bounced over the rear of the train, the report of the cannon followed instantly. Another shot followed, but it was also short, and the last wagon turned the shoulder of the hill into the gorge of the creek as the ball bounced along up the Gauley Valley.[18]

The second day's battle was then begun in earnest, continuing on and off all day. A steady rain fell throughout the day, adding additional discomfort to the troops. The Confederates succeeded in sinking a small boat near the ferry, and made matters so hot for members of the 11th Ohio that they abandoned their camp. Sharpshooters on both sides of the river repeated their performance of the previous day, with no deaths reported on either side. Again, nightfall put an end to the fighting, and both sides made plans for the third day of activity.[19]

General Rosecrans learned from scouting parties that General Lee had left Sewell Mountain, and that most of the force that had been there was now gone. Realizing that General Floyd was unsupported in this attack, it seemed possible to entrap him, and Rosecrans quickly developed a plan of action. Due to the fact that the rivers were swollen from recent rains, it would be several days before a crossing in force could be made.

Shortly after daylight of November 3, the battle commenced again on the same plan. The first shot fired by the Rebels struck near a large pot of pancake batter that was being prepared by the cook of Company C, 11th Ohio. Dirt and rocks were thrown everywhere, utterly spoiling the batter, at which time the defiant cook shouted across the river, "Fire away you damn Rebels but don't sprinkle dirt in my batter."[20]

The fighting continued and at noon word came that the Confederates had killed one man and wounded another. After considerable effort the bodies were brought off and taken to a position of safety.[21] General Benham, with 1,500 men, was posted at a point opposite the mouth of Loup Creek, some five miles from General Çox. Everyone anxiously awaited Benham's crossing, which was all the more needed as casualties began to mount for the Yankees.

On the fourth day of the engagement General Floyd's brigade commanders submitted to him a petition which they had all signed. They asked for an immediate withdrawal from Cotton Hill, as they were convinced their position could be easily flanked, and pointed out to Floyd that he was hauling provisions 100 miles and the roads were such that only one-half loads could be carried per trip. Colonel Heth, commanding the First Brigade, tried to convince Floyd that Rosecrans could cross the river and drive him away whenever he desired; to which General Floyd replied, "Let him dare cross the river and not a damn Yankee will ever recross it." Not wanting to give up so easily, Colonel Heth replied, "General Rosecrans will capture you and your entire command." To this Floyd said, "No sir, I will die before I will be captured. Do you know what they say they will do with me if captured? They say they will put me in an iron cage, haul me around their damn country as if I were a wild beast."[22]

General Floyd was firmly against withdrawal and decided instead to increase his harassment of the enemy. He ordered a detachment of troops from the 22nd and 36th Virginia Infantry, who were posted at Camp Dickerson, to mount a cannon on the heights opposite Hawks Nest, and begin shelling the camps of the 28th and 9th Ohio Infantry. This firing was in addition to that at Gauley Bridge. The artillery fire here was also brisk, but little damage was done. A ball or shell would hardly land before the Ohio boys would run with picks to dig it up as a trophy. Observing all this was six-year-old A. W. Hamilton, whose family lived in a two-story log house in the vicinity of Lovers Leap. The home was quite near one of the Ohio camps and the Rebels aimed their artillery at some caissons parked in the Hamilton yard. The boy and his family remained in the center of their home for two days and no real harm was caused by the bombardment.[23]

A Confederate soldier with the 45th Virginia Infantry, wrote to his sister on November 5, from Cotton Hill:

> Well you know when the militia had possession of this place, that all they wanted to drive the yankees out of Gauley Bridge, was one or two cannon. We have several cannons and I believe we have planted them as near the bridge as they would have planted them, and have been shooting at them for four days, and they run up and down the road and make fun of us. Aint this too bad? We will pay them for it if we get the chance. I have been where I could see all that was going on, as our company was stationed at the cannons, and I do not believe that we have killed more than five or six, if any....when we got ready and began to fire at them, we found that they had as good position as we had, and about twice as many cannons. So you see that they will hold their position if

they want to, and we will hold ours and here we will be cannonading for no telling how long.[24]

The action at Gauley Bridge on the fourth day was similar to that of the previous three days. On Wednesday, November 6, the Union troops were much relieved by the arrival of six long-range parrot guns, which had been sent up from Charleston. Lieutenant Dryden, of the 1st Kentucky Infantry, had two of his guns pulled up the steep mountain side to an elevation commanding the hills on the other side. It was nearly dark when this work was completed and the Yankees had to wait until morning to see the result of their labors.

Shortly after daylight the next morning the Rebels opened fire. Within a few minutes the Federal parrot guns had been aimed and were cut loose on the Rebel position. The first shots seemed to stun the Confederates, who ceased firing briefly and then began again. Soon, the long-range guns of General Rosecrans were making matters too hot for the short-range guns of General Floyd, and the Rebels withdrew. With the exception of light skirmishing between the sharpshooters, the Confederate siege of Gauley Bridge was over. Having been under fire the better part of seven days, the Yankee troops were much relieved to end the contest. As a result of the fighting, the Union listed two men killed and eight wounded. No returns were given for the Confederates but it is safe to assume they at least suffered a few wounded.[25]

As the siege was concluded a soldier with the 13th Georgia Infantry wrote home. He wrote:

November 7th & 8th, 1861
Miss Mary Hodnett,

> Dear sister I now take my pen in hand to yours which I received yesterday, and read them with greatest pleasure. I was truly glad to her that you are all O.K. Mary I am as well as you ever saw me I reckon. Well Mary I take a new start this morning, it was so dark that I could not see the lines, I would not have commenced last night but Mr. Cutright said that he should start home this morning but declined till evening. Mary I have no news of importance to rite this morning, we have been fighting 4 or 5 days up here with the cannons but has not hurt anything much. The Yankees is on one side of the river and we on the other side. They are fixing up for winter there but I dont think that we can stay here. Provisions is so far off and they have the Canoy Valey to get their provisions to say the least of it. I cant tell anything about the war only as it comes. The officers dont know anything about it. Well to change the subject, you ask me to tell you something about our cooking. We are getting so we can cook finely. I can cook as nice biscuits as you can I expect, though we dont have anything to cook but biscuit and beef. We have got in good practice cooking of that, it is not hard to cook. Well enough of that, you ask me to send you something of the groth of this mountain, but Mr. Cutrights leaving so much sooner than I expected, I hadnt time to

Confederate artillery firing at Gauley Bridge from Cotton Hill.
COURTESY SWV

General Henry W. Benham. General Rosecrans blamed him for allowing General Floyd's army to escape Cotton Hill.
COURTESY U.S. ARMY MILITARY HISTORY INSTITUTE

get anything more than greens there. Well Mary I must hasten on, I received my clothes by Mr. Cutright and I was truly glad to get them for I was nearly without. I have never got the others yet but Lieut. Cutright has gone after them. Now all you liked of sending me enough was a pare of shoes, I am nearly barefooted, but I will get a pare in a few days now.

Well I will pass on, tell the girls and all I will not forget them for those chestnuts and china that they sent me. Tell them all howdy and I want to see them and all of you. I hope that I can get to see you all between now and Spring, I want to get a furlo between now and Christmas if I can. Well I must bring these lines to a close, tell father I have not drawn any wages yet but we may this evening, and if he needs it I will send it to him if I cant go myself. Nothing more at present, only I remain your brother till death.

> John W. Hodnett[26]
> Cotton Hill, Fayette Co.

The fact that Confederate forces were able to remain on Cotton Hill for an extended period of time has been a point of controversy ever since the war. General Benham, with three thousand men, had crossed the Kanawha River at Loup Creek, prior to the end of the fighting. His position there was not discovered by the Confederates, and it is likely he could have surprised their force easily by an advance along Big Falls Creek.

General Rosecrans advised Benham against an advance until he had received reenforcements, which would be sent across the river at two points, once the swollen conditions of the river could be overcome. It was certainly not the fault of General Benham that these plans took several days to implement. If there had not been so much rain, and the rivers and roads been in better condition, Floyd would not have been able to act with impunity, as he had since his occupation of Cotton Hill in October.

While there was a level of inertia on the part of Benham, it was more the case during the Confederate retreat from the area than it had been during the siege. General Rosecrans would eventually call for a court martial of Benham for allowing the escape of Floyd's army. The fact that these two generals did not get along probably contributed to Rosecrans's decision. Though charges were filed and Benham's name and reputation smeared, no action was ever taken, and the incident passed into history.

When General Benham wrote an account of these events in 1873, he offered a strange but not entirely dismissable reason for Rosecrans's charges:

... I began to urgently ask for permission to attack Floyd, in his encampment, some early morning, if I could have 1,000 additional men, to make my force more nearly equal to his own. Or if I could not have the men, I asked authority to attack with my own command only, though less by 1,000 to 2,000, than I suposed he had. About this time, the sixth or seventh of November, there reached me a second telegram from General Lauder, explaining the orders issued for my detachment, and soon after, on the seventh or eighth, the New York times reached my camp, having in it a short note, stating that Mr. Lincoln had said, "If General Benham captured Floyd, he would make him a major general." (as many months after I learned the President had actually said, to one of the editors) But this unfortunate newspaper note, defeated its object; if there had been as we had hoped, such a chance in store for us; as to be permitted to attack Floyd. For myself and my officers at once recognized, as I stated to them, that "my cake was dough" that such a chance of promotion over Rosecrans he would never permit, with all his previous jealousies. And my most urgent appeals for permission to attack for several days after, were met with neglect.[27]

So the mystery remains, just as it has all these years. Although Rosecrans was commander, he escaped most of the blame for the failed campaign at Cotton Hill by shifting much of it to Benham. It is certainly by Rosecrans's neglect that Cotton Hill was not occupied by Federal forces early in October. Such a move would have prevented the advance to the area by Floyd's army, and could have been accomplished in the days immediately after their return from Sewell Mountain.

General William S. Rosecrans, commander of Union forces in Fayette County during the Sewell Mountain & Cotton Hill campaigns.
COURTESY MASSACHUSETTS COMMANDERY MILITARY ORDER OF THE LOYAL LEGION AND U.S. ARMY MILITARY HISTORY INSTITUTE

■ *Chapter Six*

The Long Arm of Lincoln

General Floyd was not aware that three thousand enemy troops under command of General Benham had crossed the Kanawha River on November 6, and had deployed for several miles up Loup Creek and along the Kanawha. Several brief skirmishes had occurred along the Cotton Hill road during the early days of November, but since these were attacks by small groups of the enemy, Floyd did not suspect a large movement at this time.

In contrast to Floyd's attitude, these skirmishes further convinced his officers that any plan to remain in their present position was hazardous. Their failure to convince Floyd that a withdrawal was necessary, further demoralized officers and men alike. The season was now advanced, and cold, damp weather, had set in. Disease and lack of provisions had taken a drastic toll on the Confederate forces, and prompted one of Floyd's officers to write the Confederate Secretary of War, on November 10. He wrote: "This army is utterly demoralized, or, if this term is too strong, it is the most disquieted collection of men I have ever known massed together. They want to go back to some point to winter nearer to provisions for men and horses."[1]

Colonel C. Q. Tompkins, one of Floyd's best commanders, resigned November 6, and his good friend, Isaac Noyes Smith, wrote to him on November 10 from his camp at the eastern foot of Cotton Hill:

> I write to speak of the regiment (22nd Virginia Infantry) The impositions upon it have been continued and most serious — our little handful of men have been made to furnish 95 men a day for duty until yesterday and today — yesterday 60 and today 20 — the effect of this heavy drain has been apparent in the morning reports — on Wednesday we had a total for duty of 296 — on Thursday 273 — on Friday 269 — on Saturday 244 & today 202 — a loss of 94 men in four days. In the last detail we sent three men barefoot & shod some of the others by taking shoes from the sick men and putting them on the men sent out for picket. Beside our picket on the hill we furnished 50 men and 2 officers to guard the Jackson battery — men from this detail who reached camp at 8 or 9 oclock p.m. have been taken for the same purpose at 3 oclock p.m. the following day & started off to the battery.... I have computed the duty done by the regiment in the 20 days of our stay here from Monday Oct. 21st. & we have averaged 75 men each day & so as to be strictly within bounds say 70 men — the character of the duty you well know — Of the 202 men reported above a number are unfit for duty for want of shoes. The whole

regiment properly is unfit for duty for want of clothes. An additional injury has been done the poor fellows—they have been more or less elated even amidst these hardships by the hope that they would receive their pay for the last four months & with that view most strenuous efforts were made to complete their rolls—as the labor was just about complete. . . . Just learn of some excitement in camp, some Yankees are said to be crossing at Montgomerys Ferry—I shall be forced to close as we may expect orders at any moment.[2]

    The information Isaac Smith received was correct. On the morning of November 10, General Cox ordered Colonel Devilliers, with 200 men of the 11th Ohio, to cross New River at a ferry which had been rigged just above the mouth of Gauley, and Lieut. Colonel Enyart, with 200 men of the 1st Kentucky, to cross at Montgomerys Ferry, and occupy as far as possible the Fayette road. At noon Devilliers's men drove in the Rebel pickets from the face of Cotton Hill, and after an extremely strenuous climb over steep and rocky hills, planted their flag in the Rebel breastworks. A lively skirmish was kept up for several hours, the lines forming on the Blake farm, about one-half way up Cotton Hill, opposite Cane Branch, which is about two miles east of Gauley Bridge. A long extended line of sentinels was required by the Yankees to guard the whole length of the ridge, which extended from above Blake's to the heights opposite Gauley Bridge.[3]

Burwell Huddleston, 22nd Virginia Infantry. He fought throughout the war and was killed by the kick of a horse in 1873. Before the war he lived on Falls Branch, Cotton Hill.
COURTESY TERRY LOWRY, CHARLESTON, WV

Shortly after dark six companies of the 2nd Kentucky Infantry crossed the river to reenforce Devilliers. The Confederates were also reenforced and at 9 p.m. they attacked the left wing of the 11th Ohio, under Major Coleman, and drove it back from Blake's farm about a quarter of a mile. The fighting was sometimes heavy, though it was difficult to maneuver in the mountain darkness. The bright flash of fire from muskets exposed enemy positions that otherwise would have been almost impossible to determine. The Federals were soon reenforced near Blake's and quickly recaptured their former position, driving the Rebels to the crest of Cotton Hill. The Confederates made a succession of attacks upon the Yankees who were pushing their way up the crest along the whole line from Blake's to the Kanawha. Brisk skirmishing was kept up until after midnight, when the Federals succeeded in securing the entire ridge above Blake's.[4]

Returning to camp in the darkness, members of the 51st Virginia Infantry stumbled onto pickets of their own regiment. After they began to fire upon each other, the 20th Mississippi Infantry became involved. Before the affair ended, one man from the 51st was dead and three wounded. Several Mississippians were also wounded in the melee. These were the only Confederate casualties from the fighting of the day and night of the tenth. The Federals suffered three killed, two wounded, and five taken prisoner, as well as the loss of seven muskets, with accouterments.[5]

General Floyd's decision had now been made for him and a retreat was planned. On the morning of the eleventh, Colonel Devilliers, with the 11th Ohio and 2nd Kentucky troops, pushed forward and drove the Rebels from the heights of Cotton Hill, and reached a position where they could see the Confederate baggage train moving along the Fayette turnpike from Huddleston's. At 9 a.m. the Confederates were also observed breaking camp at Laurel Creek, in present-day Beckwith, and retiring towards Dickerson's. Upon reaching Dickerson's it was decided to add some heavy timber to the breastworks already at that point. A cold rain poured all day and it was with great difficulty that fires were kept. Fear of capture multiplied the anxieties of Floyd's troops and much confusion and excitement was prevalent.

The next day, the 12th, General Benham was ordered to advance from his position at the mouth of Loup Creek, and attack Floyd from Cotton Hill. General Rosecrans hoped to attack Floyd from front and rear simultaneously, by sending 1,000 men up the left fork of Loup Creek, so as to attack Floyd from Cassidy's Mill. Benham advanced with 1,640 men and six cannons. The Confederates at the eastern base of Cotton Hill prepared an ambush for the Yankees in the ravine of Laurel Creek, at present-day Beckwith. Skirmishing was kept up all afternoon on the twelfth, and General Floyd received word of Benham's movements from a civilian, as described by Colonel Henry Heth: "I was skirmishing with Rosecrans's advance when Floyd, much excited, rode up to me and said a country girl had just ridden in and informed him of the approach of Benham's force. 'By God, they will get in my rear; I shall be attacked in front and rear; what must I do?' I think visions of the iron cage were very vivid then. I said, General, leave me with one regiment, a battery of artillery, and a dozen cavalry. Take the rest of your command, and if the mouth of the sac is occupied, cut your way through."[6]

General Floyd took this advice and withdrew all but a few of his troops to Camp Dickerson. Those that remained ambushed the Yankee advance on Laurel Creek at three o'clock. Colonel William S. Smith, of the 13th Ohio Infantry, described the ambush. He wrote: "... Cpt. Careys company of the 12th Ohio and Cpt. Beaches company of the 13th Ohio, were ordered to make a reconnaissance of the Laurel Creek ravine just in advance of us and through which our road lay for the distance of about one half a mile. These companies had but fairly entered the ravine when they came upon a strong outpost of the enemy lying in ambush. A sharp skirmish ensued, which was poured in upon them at short range." General Benham ordered his troops to retire a short distance and encamp. His forces had suffered two killed and three wounded in the ambush; the Confederates had two killed and seven wounded.[7]

That night the Confederates began a somewhat disorderly retreat from Fayette County. Due to their haste and a shortage of wagons, a great deal of valuable material was destroyed. The torch was applied to about three hundred tents, several bales of new blankets and overcoats, mess chests, and camp equipage of all kinds. Flour barrels were burst and the contents scattered upon the ground, as well as all kinds of provisions, all to keep the enemy from getting them.[8]

General Benham did not send scouts ahead until 9 p.m., and several hours passed before any word reached him concerning the Confederate movement:

> I rested myself with my staff, on a few pine branches, on a shelf of a rock, of two or three yards square, overhanging the road, where my orderlies and escort, of 12 or 15 cavalry, set down, holding their horses during the whole night. During which I had but an hour or to of sleep, from the receipt of, and replies to, messages from Rosecrans and others. About two a.m. of the 13th, my scouts came in and reported having passed some two miles to the front over the fields; but without meeting any troops. They found however from the noise, as they approached the Dickerson farm, that over this ridge, there was a constant rumbling noise, as of the moving of wagons and artillery, but they could not form the slightest judgment whether the enemy were in retreat or receiving reenforcements.[9]

General Benham ordered out additional scouts, but due to a mix-up in orders they were not sent. The commander of Benham's force at Cassidys Mill also sent scouts who were in Fayetteville at 9 a.m. on November 13th, but for unknown reasons failed to inform Benham of their findings.[10] When Benham realized on the morning of the 13th that his order had not been obeyed, he personally ordered scouts to advance towards Fayetteville and report as soon as possible. These men spent all day covering a distance of two miles, from the eastern foot of Cotton Hill, to the position Floyd had held on the Dickerson farm. When they reported to Benham at 4:30 p.m., he began an immediate pursuit, but was now almost twenty-four hours behind Floyd. General Benham described their pursuit. He wrote:

> We moved on now about 5 p.m. as rapidly as possible, orders being left for the detachment coming over the mountains, to follow and join me.... we crossed the ridge of Dickersons farm and we found it to be a most strongly entrenched position. And some two miles further on, we came about dark, to the position apparently of the main camp of the Rebels, seemingly just deserted, and with the remains of hundreds of tents and camp equipage destroyed in their fires, with small arms, ammunition, and private baggage, even to personal arms, swords and etc., strewn around, as if left in the greatest haste.[11]

General Benham determined to rest his command at this point, which was then the farm of Levi Jones. His force was soon reenforced by the 47th and 7th Ohio, swelling his command at that position to 2,700 men. At eleven p.m. he resumed the march and reached Fayetteville around midnight. An unknown member of the 13th Ohio Infantry described the situation at Fayetteville: "Here quietness reigned supreme, not even a dog howl greeted us, and, in short, this deserted village presented inhospitality in all its phases. Anyone who has experienced a night march can appreciate our feelings, when moving in silence over an unknown road, in expectation of meeting the enemy at any moment. The dull, heavy, monotonous tread of the men, and the sound of the horse hoof on the road made the most self possessed of us reflect upon what may be our fate. . . . After half an hour we passed through the village with an involuntary desire to reduce it to ashes."[12]

The Federals did not reduce Fayetteville to ashes, though they had attempted to do so on October 19. The pursuit continued, the troops marching through rain and fog all night, they reached Hawkins farm, five miles south of Fayetteville, at four a.m. of November 14. They left this farm at 7 a.m., and marched just four miles, when they encountered Confederate scouts at the rear of Floyd's army. This position was about one mile from McCoys Mill, present-day Glen Jean. Federal skirmishers of the 13th Ohio were in advance of the main column on both sides of the road. As the Yankees came to a bend in the road Private Seig of Company F, crawled to a position where he could view the road in their front. To his surprise he saw a Rebel cavalry party advancing in his direction.[13] These men were under command of Colonel George Croghan, Floyd's cavalry commander. Confederate Colonel B. Estvan described the event. He wrote: "A team of carts laden with provisions had been delayed, and ran the risk of falling into the hands of the enemy, who were almost at our heels, and already sending a few bullets among us to hasten our flight. When Col. Croghan, followed by twenty five of his lancers, dashed down the road to check the enemy, with the view to save the carts, but he had scarcely reached the latter, when two bullets brought him to the ground."[14]

A Federal account of Colonel Croghan's death later appeared in the *Cincinnati Gazette*: "A volley was instantly opened upon the enemy, who were taken completely by surprise. At the first fire several saddles emptied, and Col. Croghan fell mortally wounded in the abdomen. The Rebels, though surprised, showed fight and retired slowly, firing as they went; but our men having possession of the elevated ground on both sides, exposed them to a galling cross-fire, and forced them back. Colonel Croghans father and General Benham were old acquaintances."[15]

Croghan was mortally wounded and asked to see General Benham, who was then nearing the scene, having heard the firing in his front. Benham later described their meeting. He wrote: "I of course stopped and dismounted for a few minutes to see him, and found that he was the son of our late Inspector General, Colonel Croghan. He had been shot through the waist, twice through the sword belt as I recollect. He had a memorandum of his wishes made, as he conversed without great pain, and requested that I would state to his friends that he died like a brave soldier; and I left the house, commending him to the care of the occupants, a Miss Fanny Hill, and a deaf mute sister."[16]

General Floyd's forces were camped in the vicinity of McCoys Mill, about one mile distant from the Hill house. At this time no one in camp realized that the Yankees had closed to within such close proximity. Suddenly, Croghan's Lancers rushed into camp in search of Floyd and announced Benham's presence in their rear. This shocking news was almost more than the weary troops could bear, as described by an unknown Confederate writer: ". . . all broke off in a wild run, some so frightened that they threw away their knapsacks and all they had, except gun and knife to protect themselves. It required great effort on the part of the officers, who were somewhat cool, to prevent a perfect rout. The road was so bad and muddy, that the brigade could not march more than eight miles a day. There had been so much rain and waggoning along the road that it was a perfect mire, about half a leg deep, and all had to wade right through it."[17]

General Floyd's main force began a rapid retreat, while two

regiments of infantry and a detachment of cavalry remained behind to slow the enemy's progress. Benham's advance reached McCoys Mill at 12:20 p.m. and discovered the Rebels posted behind a long ridge which overlooked the mill. As the Yankees came into range the Confederates opened a heavy skirmishing fire on them and halted their advance. General Benham at once ordered the 7th and 36th Ohio to take position on a ridge at the right of the road. The 13th Ohio moved to the left and Captain Schneider's artillery battery was brought to the extreme front, as described by a member of the 7th Ohio Infantry: "We now poured it in hot and heavy, and they scattered in all directions. All this time our troops on the right were firing whenever they had a fair chance and advancing all the time. The 13th took up a position on their extreme left, threatening to get in their rear. The Rebels, finding the climate becoming too warm for even their Southern constitutions, and the thermometer constantly rising, fled in disorder, dashing down through a corn field, our men popping away at them in the most lively and pleasant manner."[18]

As the Rebels retreated Benham's troops advanced cautiously and took possession of the ground Floyd had occupied. A short rest was ordered, after which the pursuit resumed in a heavy downpour of rain. After advancing but a few miles a halt was ordered at the Keton farm, fifteen miles from Fayetteville. The Confederates continued their flight, and at ten p.m., a dispatch reached General Benham, from General Schenck, ordering him to return to Fayetteville as soon as possible. Due to the fact that many of his men were still without tents, and a heavy storm was then raging, Benham decided to retrace their steps that very night. They commenced their return at 1:00 a.m., reaching McCoy's at 4:00 a.m., where they rested until 6 a.m. of the 15th, and reached Fayetteville at 1 p.m.[19]

General Floyd could not know that the enemy pursuit had been terminated. He ordered the 22nd and 36th Virginia Infantry to take a position several miles in rear, where they were to fight the Yankees on the morning of November 15th. His main force continued their march, stopping eight miles from Raleigh Court House. His rear guard remained in position until 9 a.m. of the 15th, and seeing that no Yankees were in the area, they returned to camp. Major Isaac Noyes Smith, of the 22nd Virginia Infantry, was among the rear guard troops and later wrote an account of their march. He wrote:

> On this march we saw the true character of our retreat. The road and roadside were strewn with articles of every description—tents, boxes, guns, knapsacks, broken harness, cooking utensils and dishes. Wagons were left fast in the mud with their loads untouched; at one place twelve wagons were left, most of them turned upside down, and at the same

place a number of horses had drowned in the mud, that is had sunk beyond hope of extrication and been shot. At some places mud holes were filled up with tents to make a passage for the wagons. All this was a perfect harvest to the 22nd., being in front of the regiment we noticed there was only a fragment following up, and halted for the stragglers to catch up; waited a long time and they did not appear—went back and found the fellows resting behind. The rascals claimed to be tired down as a reason for their halt, and no wonder, for when I started them again I found about every other man loaded down with flour, frying pans, buckets, mess kettles and such things. I found myself, eleven good percussion muskets, and gave to the men, but made them carry their flintlocks also.[20]

General Floyd's command withdrew to the area of Dublin Depot, Virginia. Several days after he left Fayette County he received a dispatch from the Confederate Government at Richmond, asking him to hold Cotton Hill:

War Department C. S. A.
Richmond, November 15, 1861

Brig. Gen. John B. Floyd, Commanding Army of Kanawha:

Sir: I have hitherto refrained from replying to your several letters in relation to your proposed movements during the coming winter, because it was necessary first to ascertain what force would be under your command, and whether such force could reasonably be expected to succeed in any offensive operation. I have at last succeeded in sending to your aid three fine regiments, that will be with you before your receipt of this letter, one under Col. Starke, and two Tennessee regiments under Brigadier General Donelson. With this force the president is satisfied you ought to be able to hold your position at Cotton Mountain, and he hopes you will not fail to do so, as it is very obvious that on your abandonment of so important a point the enemy, now taught by experience, will not fail to seize it. Hardships and exposure will undoubtedly be suffered by our troops, but this is war, and we cannot hope to conquer our liberties or secure our rights by ease and comfort. We cannot believe that our gallant and determined citizen soldiers will shrink from a campaign the result of which must be to drive the enemy outside of our borders and to secure for us the possession of a valley of such vast importance as that of the Kanawha, at the present critical juncture. I therefore hope that you will not feel compelled to abandon Cotton Mountain in order to fall back on Raleigh Court House, or any other point, until you have forced the enemy to abandon their camp at the junction of the Gauley and Kanawha. I have sent you a rifled twelve-pounder within the last few days, and will send you another in a few days more. I am very sorry we have no 24-pounder howitzers. Do your best to keep your road to Newbern in transitable order, and supplies shall not fail you.
I am, your obedient servant,

J. P. Benjamin,
Acting Secretary of War.[21]

Obviously, it was too late for General Floyd to hold Cotton Hill, or any other point in Fayette County. His withdrawal signaled the end of Confederate efforts in this region for 1861. On his very long and arduous retreat his forces lost about eight men killed. Not all were victim to Yankee bullets; some died from exposure, and many others became sick and were hospitalized. The death of Colonel Croghan was another severe blow to Confederate morale, at a time when it would seem that things could not get worse for Floyd's command. Colonel Croghan was the fourth Confederate officer to be killed in Fayette County.

With no organized Rebel force remaining in Fayette County, the Federal Government focused its attention on more active theaters of war. On November 19, eight regiments were taken from Rosecrans's command and sent into Kentucky.[22] One week later Rosecrans had General Benham arrested and charged with unofficer-like conduct in allowing General Floyd to escape. On November 30, four infantry regiments and three batteries of artillery were taken from Rosecrans, some going into Kentucky, others were sent to Romney and Washington D.C. With his remaining forces Rosecrans made plans for the winter. Schenck's brigade was posted at Fayetteville; the 47th Ohio with three cannons at Gauley Mountain; the 28th Ohio at Gauley Bridge; the 36th Ohio at Cross Lanes and Summersville; the 37th Ohio at Cannelton and Loup Creek; the 44th Ohio at Camp Piatt near Charleston; the Cox brigade at Charleston and along the Kanawha River; the 8th Virginia at Buffalo; the 4th Virginia at Point Pleasant; the 34th Ohio at Barboursville and Mud River; the 2nd Virginia Cavalry at Guyandotte; and the 5th Virginia at Ceredo.[23]

As General Rosecrans was making these troop placements, General Floyd was resting his command near Dublin Station, Virginia, on the Virginia and Tennessee Railroad. After a rest and reorganization, all of his command, except the 22nd Virginia Infantry, was transferred to Bowling Green, Kentucky, to serve under General Albert Sidney Johnston. The 22nd Virginia was sent to Lewisburg to replace the Wise Legion, which under the command of Colonel J. Lucias Davis had been recalled to Richmond.

On December 5, Mrs. Ellen Wilkins Tompkins, wife of Colonel C. Q. Tompkins, abandoned her Gauley Mountain estate and traveled to Richmond to join her husband who had gone there after resigning from the army in November. Before leaving Mrs. Tompkins wrote to her sister expressing her feelings about the move. She wrote:

> I shall go to Richmond and no matter what happens stand my ground. I have been very busy packing, arranging things, and am filled with disgust to see this beautiful place torn to pieces by the soldiers, when I remember the cost of money and trouble to build all the houses, etc. It is

a perfect desolation now. Fences are all gone. The fields set in clover are hard as roads from the encampments. What a pity we ever saw the place. I cannot remain here this winter to save it, as in the Spring the fighting will begin again. I have had as many cannon balls roaring around me as I wish to hear. There is no hope of peace for years. The South will not accept the terms offered by the North. The wind is roaring, rain, hail, and snow falling. I dread this journey.

In order to travel freely along the turnpike it was necessary for Mrs. Tompkins to obtain a pass:

> Head Quarters, 47th Regiment Ohio Volunteers U.S.A.
> Camp Gauley Mount, December 3rd, 1861
>
> To all whom it may concern.
> The persons herein named to wit:
> Mrs. Ellen Tompkins with her two sons, Joseph and Willie Tompkins, also three colored women, five colored children are permitted to pass free and unmolested through the Union pickets on the road leading to Meadow Bluff.
>
> Officers and Guards will respect this.
> By Order of General Rosecrans
>
> Colonel F. Poschner
Commanding Post[24]

On December 15 Major E. B. Andrews of the 36th Ohio proceeded with 150 men by the Wilderness Road to Meadow Bluff, Greenbrier County. He found the Rebel encampment there deserted and burned 110 log huts, and some tents. They also captured two noted guerrillas, 21 rifles and guns, 21 mules and horses, 95 cattle, and 200 sheep. They had a brief skirmish with Rebel cavalry and returned on December 21st.[25]

On December 20 the forces stationed at Fayetteville began constructing two forts. One was on a hill northeast of town, the other on a hill southwest of town. There were 1,430 men camped at Fayetteville, consisting of: 23rd Ohio, 550; 26th Ohio, 600; 30th Ohio, 200; Artillery, 40; cavalry, 40. The forts were completed on December 28, and the following day, Sunday, five infantry companies marched from Fayetteville to Raleigh Court House, so as to maintain control of that community and the various approaches to Fayetteville.[26]

The West Virginia campaigns of 1861 were now over. General Rosecrans had begun the campaign in August with the announced purpose of marching to Wytheville, Virginia, and on into the Holston Valley. General McClellan had cherished the idea of making the Kanawha line the base of operations into the same region. It was easy to sweep a hand over a few inches of map, showing nothing of the topography, and to say, "We will march from here to here." The almost wilderness nature of the country, with its weary miles of steep mountain roads that became

"Gauley Mount" home of Colonel C. Q. Tompkins of the Rebel army.
COURTESY AUBREY MUSICK, GAULEY BRIDGE, WV

impassable in rainy weather, and the total absence of forage for animals, were elements which the Federal commanders greatly underestimated.

Instead of reaching Wytheville, Rosecrans found that he could not supply his little army, even at Big Sewell Mountain. It was not General Floyd's army, but the physical obstacles presented by the country, that chained him to Gauley Bridge. The Southern troops could not hold Western Virginia for a number of reasons. The Allegheny Mountains divided West Virginia from the main Confederate army. A large percentage of the populace of the trans-Allegheny region were sympathetic to the Union. For the most part, the generals who led the Federal forces were competent, while Wise and Floyd were generals in name only. The Federal leaders were united in purpose and command, while the two jealous and quarreling ex-governors let their opportunities slip through their fingers.

The spectacular campaigns of the Army of the Potomac in the East and of General Grant in the West relegated the West Virginia campaigns of 1861 into obscurity. The Southern Confederacy lost the mineral resources (especially salt) of the Kanawha region. The Baltimore and Ohio Railroad remained in Federal hands. More than that, a new state appeared on the horizon, for these military operations assured the addition of a new star to the flag. The Union success in West Virginia was undoubtedly a welcomed relief to President Lincoln, who hungered for, but had not obtained, similar results in eastern Virginia.

While at Dublin Depot, Virginia, General Floyd issued an optimistic and congratulatory order to his troops:

General Orders,   Hdqrs. Army of the Kanawha
No. -   Camp near Dublin Depot, December 1861

SOLDIERS OF THE ARMY OF THE KANAWHA:

The campaign in the western portion of this state is now, as far as you are concerned, ended. At its close you can review it with pride and satisfaction. You first encountered the enemy five months since, on his unobstructed march into the interior of the state. From that time until recalled from the field you were engaged in perpetual warfare with him. Hard contested battles and skirmishes were matters of almost daily occurrence. Nor is it to be forgotten that laborious and arduous marches by day and by night were necessary, not only as furnishing you the opportunity of fighting these, but of baffling the foe at different points upon his march of invasion. And it is a fact which entitles you to the warm congratulations of your general, and to the thanks and gratitude of your country, that in the midst of the trying scenes through which you have passed you have proved yourselves men and patriots, who, undaunted by superior numbers, have engaged the foe, beaten him in the field, and baffled and frustrated him in his plans to surprise you. On all occasions, under all circumstances, your patriotism and courage have never failed nor forsaken you. With inadequate transportation, often illy clad, and with less than a full allowance of provisions, no private has ever uttered a complaint to his general. This fact was grateful to his feelings, and if your hardships have not been removed or alleviated by him, it has been because of his inability to do so. But your exemplary and patriotic conduct has not passed unobserved and unappreciated by the Government in whose cause we are all enlisted. It is an acknowledged fact that you have made fewer claims and imposed less trouble upon it than any army in the field, content to dare and to do as became true soldiers and patriots with the means at your command. Now, at the close of your laborious and eventful campaign, when you may have looked forward to a season of rest, your country has bestowed upon you the distinguished compliment of calling you to another field of action. That you will freely respond to this call your past services, so cheerfully rendered, furnish the amplest assurance. Kentucky in her hour of peril appeals to Virginia, her mother, and to her sisters for succor. This appeal is not unheeded by their gallant sons. The foot of the oppressor is upon her. Trusting in the cause of justice we go to her relief, and with the help of him who is its author we will do our part in hurling back and chastising the oppressor who is desecrating her soil. Soldiers, your country, your friends whom you leave behind, will expect you in your new field of labor to do your duty. Remember that the eyes of the country are upon you, and that upon your action in part depends the result of the greatest struggle the world ever saw, involving not only your freedom, your property, and your lives, but the fate of political liberty everywhere. Remembering this, and relying upon Him who controls the destinies of nations as of individuals, you need not fear the result.

By order of Brig. Gen. John B. Floyd:

H. B. Davidson,
Major and Assistant Adjutant-General[27]

Confederate belt buckles found by the author in Fayette County.

Louisiana "Pelican" belt buckle found in Fayette County by Sam Pittman of Charleston, WV.

Rare Ohio State Seal breast plate found by the author in Fayette County.

Rare Ohio Volunteer Militia belt buckle found by the author in Fayette County.

Artillery shells and bayonet of the type used in Fayette County.

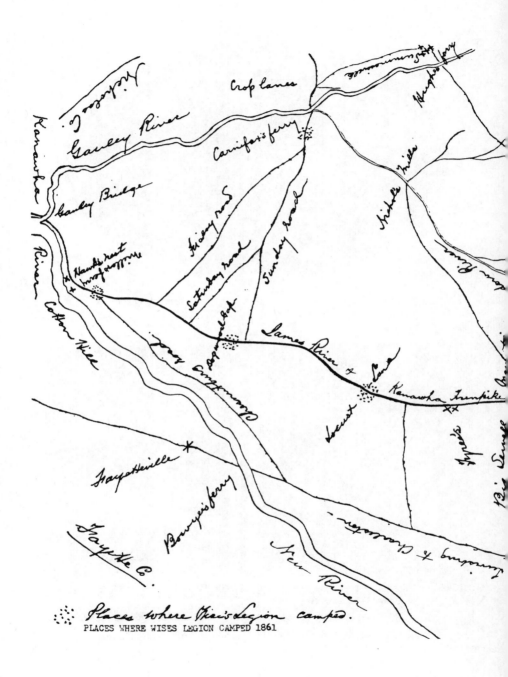
PLACES WHERE WISES LEGION CAMPED 1861

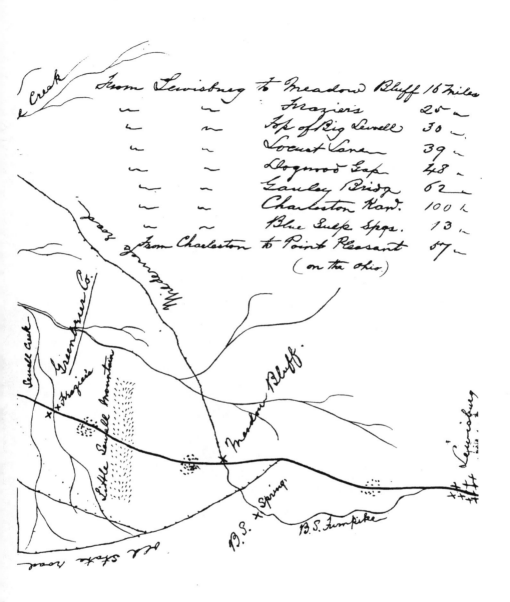

Manuscript map of area between Gauley Bridge & Lewisburg, WV, drawn in 1861.
COURTESY VIRGINIA HISTORICAL SOCIETY

1862 picture of the wire suspension bridge which spanned the Gauley River until destroyed by Union troops in September 1862.
COURTESY AUBREY MUSICK, GAULEY BRIDGE, WV

■ *Chapter Seven*

Civilian Prisoners and Refugees

Compared to the events of 1861 the period between January and August 1862 was relatively peaceful in Fayette County. Rather than organized Rebel activity, Union troops had to contend with occasional attacks by Confederate militia or by partisan rangers, organized by Captain William D. Thurmond. The activities of Southern "bushwackers" also seriously affected the Federal Government's ability to maintain firm control of this area. "Bushwacker," was a term applied to marauding bands of men who were not actually enlisted in the Southern army, but who sought by whatever means, to disrupt and destroy Union efforts in western Virginia and many other areas of the South. In fairness, it should also be noted that Union bushwackers were prevalent in some areas and equally despised by their enemies. It has been said that the ubiquitous bushwackers knew every isolated path and byway and had a virtually unassailable sanctuary in the sparsely settled interior of western Virginia.[1]

During the early months of 1862 a strong Union force held Fayette County and the Kanawha Valley generally. It expended much energy in fortifying Gauley Bridge and Fayetteville against attack. At Gauley bridge were 889 men of the 28th Ohio Infantry and 88 men of the 1st Illinois Dragoons; at Gauley Mount or Tompkins farm, 794 men of the 47th Ohio Infantry, and 107 men from a company of Ohio artillery; and at Fayetteville, 670 men of the 30th Ohio Infantry, and 377 from the 23rd Ohio Infantry, along with 98 men of the Ohio artillery, for a total of 3,023 Federal troops in Fayette County.[2]

The artillery consisted of McMullin's battery at Fayetteville with two bronze 6-pounder smooth-bores, two 10-pounder Parrot guns, and four 12-pounder mountain howitzers. Simmond's battery was at Gauley Bridge and Gauley Mount. At the bridge were posted three 10-pounder Parrot guns, two bronze rifled James guns, and one smooth-bore 6-pounder captured from Wise. At Gauley Mount were six 12-pounder mountain howitzers. An additional 8,000 troops under the general command of General Cox were posted in various areas of the Kanawha Valley and adjacent country.[3]

133

On January 25 a new wire suspension bridge was completed over the Gauley River, replacing the one destroyed by General Wise the previous July. The *Cincinnati Commercial* of February 17 1862, carried the following notice concerning the new bridge:

> The Gauley bridge, burned by the Rebel General Wise, has been rebuilt by Captain E. P. Fitch, the brigade quartermaster, attached to the staff of General Cox. It was constructed in 23 working days from the date of making the contract, and was opened for travel on the first day of this month. This bridge is about 585 feet long, 10 feet wide, divided into 3 spans. The main sustaining parts are one and one quarter inch wire ropes. The roadway is of wood, and so ingeniously braced that detachments of cavalry ride over it at a charge, producing no more, or, in fact, not as much vibration, as is induced under similar circumstances on a thorough truss bridge. The 28th Regiment Ohio Volunteers, Colonel Moore, Captain Simmons battery, and Captain Shonbergs cavalry marched and countermarched across it some days since for the purpose of testing its stability. The entire 28th regiment was closely packed on one span and a half, two sections of Captain Simmons battery occupying another span at the same time. This immense load was born at a halt and in motion, portions of it marching to the music of the band at a cadence step, without producing the slightest evidence of weakness. The entire work was executed by Stone, Quigley & Burton, bridge builders of Philadelphia.[4]

In an attempt to facilitate the transportation of supplies to and from Gauley Bridge, General Cox had constructed keel boats which proved to be a valuable asset in river navigation. These boats were able to run at all times between the head of steamboat navigation and the Kanawha Falls, one and one-half miles below Gauley Bridge. They were built sixty feet long by eight feet wide and were shaped like a canoe. They had a capacity of eight tons and were pushed by a five man crew using poles. General Cox reported their use in a dispatch to General Rosecrans; "I have two running between Loup Creek and the Kanawha Falls, and two more are nearly completed. . . . These which I have built cost $250.00 apiece. A dozen of them could be procured at Cincinnati in as many days. . . . for the same amount of transportation they are cheaper than wagons, use fewer men, save wear and tear of teams and harness, and make less exposure of goods to theft and loss."[5]

Due to lower than anticipated enlistments into Confederate service from the counties of southern West Virginia, and due to a general concern for the South's ability to control the region, Confederate President Jefferson Davis declared martial law in Fayette and other counties on March 29 1862:

General Orders, No. 18

War Department
Adjt. and Insp. Generals Office
Richmond, March 29, 1862

I. The following proclamation is published for the information of all concerned:

PROCLAMATION

By virtue of the power vested in me by law to declare the suspension of the privilege of the writ of habeus corpus—I, Jefferson Davis, President of the Confederate States of America, do proclaim that martial law is hereby extended over the counties of Greenbrier, Pocahontas, Bath, Alleghany, Monroe, Mercer, Raleigh, Fayette, Nicholas, and Randolph, and I do proclaim the suspension of all civil jurisdiction (with the exception of that enabling the courts to take cognizance of the probate of wills, the administration of the estates of deceased persons, the qualification of guardians, to enter the decrees and orders for the partition and sale of property, to make orders concerning roads and bridges, to assess county levies, and to order the payment of county dues), and the suspension of the writ of habeus corpus in the counties aforesaid.

In faith whereof I have hereunto signed my name and set my seal this the 29th day of March, in the year 1862.

Jefferson Davis

II. Brig. Gen. Henry Heth is charged with the due execution of the foregoing proclamation. He will forthwith establish an efficient military police, and will enforce the following orders:

III. All distillation of spirituous liquors is positively prohibited, and the distillers will forthwith be closed. The sale of spirituous liquors of any kind is also prohibited, and establishments for the sale thereof will be closed.

IV. All persons infringing the above prohibition will suffer such punishment as shall be ordered by the sentence of a court martial; provided that no sentence to hard labor for more than one month shall be inflicted by the sentence of a regimental court martial as directed by the Sixty-seventh Article of War.

Command of the Secretary of War:

S. Cooper
Adjutant and Inspector General[6]

This declaration of martial law was just one more hardship which the civilians of southern West Virginia had to endure. They had already been subjected to numerous depredations and infringements at the hands of military forces on both sides. During the course of the war, hundreds of people were imprisoned by Union and Confederate forces.

The following pages list civilian prisoners of Fayette and adjacent counties being held or recently released in March 1862. In reading the lists it should be remembered that dozens of civilians had been jailed and released whose names do not appear here, and that numerous others were arrested before war's end.

CIVILIAN PRISONERS FROM FAYETTE & ADJACENT COUNTIES, HELD BY U.S. FORCES, MARCH 1862

NAME	RESIDENCE	CHARGE
Allen, William, Sr	Kanawha Co.	Secessionist
Allen, William, Jr	Kanawha Co.	Secessionist
Amick, Eli	Nicholas Co.	Secessionist
Burdett, John	Greenbrier Co.	Bushwacker
Cavendish, John	Fayette Co.	Secessionist
Dillon, R.H.	Kanawha Co.	Secessionist
Duffy, Peter	Fayette Co.	Secessionist
Farley, J.W.	Kanawha Co.	Secessionist
Fleshman, C.S.	Kanawha Co.	Secessionist
Flint, William	Kanawha Co.	Secessionist
Holmes, N.G.	Kanawha Co.	Aiding Rebels
Kious, J.C.	Fayette Co.	Secessionist
Lewis, John E.	Kanawha Co.	Aiding Rebels
Lewy, Green	Fayette Co.	Secessionist
Manie, A.J.	Fayette Co.	Secessionist
Martin, David	Fayette Co.	Secessionist
McClung, George	Nicholas Co.	Bushwacker
McCutcheon, Jones	Fayette Co.	Secessionist
Morris, W.B.	Nicholas Co.	Secessionist
Neal, Anderson	Fayette Co.	Secessionist
Neal, Johnathan	Fayette Co.	Secessionist
Norton, Moses	Kanawha Co.	Secessionist
Odell, W.H.	Nicholas Co.	Aiding Rebels
Parsons, C.C.	Kanawha Co.	Secessionist
Pierce, Benjamin	Raleigh Co.	Secessionist
Props, Noah W.	Nicholas Co.	Bushwacker
Rogers, Lem	Fayette Co.	Secessionist
Rogers, William	Fayette Co.	Secessionist
Rooke, E.	Kanawha Co.	Aiding Rebels
Rowsey, Kilburn	Fayette Co.	Secessionist
Schakelford, Cobb	Nicholas Co.	Secessionist
Sevey, Herman	Fayette Co.	Rank Bushwacker
Smith, Allen	Nicholas Co.	Aiding Rebels
Stevens, J.J.	Fayette Co.	Secessionist
Stollings, Granville	Kanawha Co.	Secessionist
Van Bibber, D.C.	Nicholas Co.	Secessionist
Van Bibber, J.C.	Nicholas Co.	Secessionist
Vance, George	Nicholas Co.	Secessionist
White, W.T.	Nicholas Co.	Secessionist
Windsor, Anderson	Fayette Co.	Secessionist

TOTAL: 41

LIST OF CIVILIAN PRISONERS FROM FAYETTE AND ADJACENT COUNTIES BEING HELD BY CONFEDERATE FORCES, MARCH 1862

Anderson, Peter L.—Age 40, Arrested for suspected Union sympathy. Deserted from Col. Beckleys Militia. Lives in Fayette County.

Armstrong, Stewart—Age 25, Lives in Fayette Co. Voted against secession. Willing to take oath of allegiance to the south.

Bays, Isaac—Lives in Fayette Co. No record why he was arrested. Says he is a southern man. Recommend oath of allegiance & release.

Bays, Samuel—Lives in Fayette Co. Has always been a southern rights man. Probably arrested because there was a general removal of citizens in the rear of Floyds army. Will take oath & release.

Clay, Charles—Age 63, Lives in Raleigh Co. No record why he was arrested. Says he is a strong southern rights man. He has two sons in Floyds brigade. Will take oath & release.

Combs, Jeremiah—Born in Raleigh Co. No charge sent in with him. Says he is a secessionist. Give oath & release.

Cornan, James—Lives in Nicholas Co. Says the Yankees camped near his house and came to him for corn. He traded corn for coffee with Yankees. Says he served in the militia briefly and has a brother in the Wise legion. Give oath & release.

Deekens, John—Age 49, lives in Raleigh Co. Does not know why he was arrested. Thinks it was from malice of his enemys. Has never seen a Yankee. Give oath & release.

Deekens, William—Lives in Raleigh Co. Says he is a southern man. Claims to have had no dealings with the Yankees. Give oath & release.

Dickens, Hardman—Lives in Raleigh Co. No charges brought, no proof given. Professes loyalty to the south. Give oath & release.

Eades, Stephen—Lives in Fayette Co. Voted against secession. Says he has had nothing to do with Yankees. Was in southern militia for a brief period. No witnesses against him. Give oath & release.

Fellow, Otey—Lives on Laurel Creek, Fayette Co. Arrested by the Caskie Rangers for aiding Yankees. Says he sent three sons to the southern army. Claims one son was wounded at Scary Creek. Evidence is strong against this man. Should be held as prisoner of war.

Flanagan, R. A.—Age 55, Lives in Fayette Co. Voted against the ordinance of secession. Claims to support constitution of Virginia and of the Confederate States. Was arrested on a visit to his sick son, who is in our army. Give oath & release.

Fox, George W.—Lives in Fayette Co. Does not know why he was arrested. He was taken to Floyds Gauley Camp just before battle of Carnifax Ferry. Was under guard across the river during the battle. Witnesses prove he is man of good character. Give oath & release.

Fuller, Alexander—Age 22, Lives in Fayette Co. Went to see his sick brother at Charleston, Va. Was arrested on trip back. No proof of anything. Give oath & release.

Fuller, Jesse—Age 24, Lives in Fayette Co. Arrested along with his brother, Alexander Fuller. Give oath & release.

Gesh, A. B.—Lives in Fayette Co. Arrested by Beckleys militia on suspicion. No proof of anything. Give oath & release.

Haywood, Thomas—Lives in Nicholas Co. Is a Union man but claims no complicity with the enemy. Is willing to take oath to the south.

Honaker, John—A youth, born in Fayette Co. says his father voted for secession, and he is a southern man. Says his horse was taken. He went to get him and was arrested. Give oath & release.

Hunt, George—Lives in Fayette Co. Voted against secession but has taken no part in the contest. No proof of anything. Oath & release.

Jackson, Maben—Lives in Raleigh Co. Was in Beckleys militia until his wife got sick. He was permitted to go home. Says he was home at the time the Yankees came to Raleigh Court house. He has permission signed by H. C. Richmore, to go home. Recommend his discharge.

Jarrett, Oliver—Lives in Kanawha Co. on Cabin Creek. Arrested on suspicion only. Says he has no contact with Yankees. Oath & release.

Jarrett, Seth—Lives in Kanawha Co. Brother of Oliver Jarrett. Says when Wise legion was on the Kanawha, he fixed guns and swords. Give oath & release.

Johnson, Miles—Lives in Fayette Co. Was born on Loop Creek. Was arrested by Caskies rangers on suspicion. Give oath & release.

Jones, William—Lives in Fayette Co. Keeps a tavern near Dogwood Gap. Some Yankees got dinner at his house. Claims no dealing with the Union government. Give oath & release.

Kelly, William—Lives in Fayette Co. on Laurel Creek. Arrested for suspicious activity. Witnesses against him claim he is a spy. Should be held as a prisoner of war.

Kincaid, James, Sr.—Lives in Nicholas Co. Arrested by Wise legion. Says he was once in Floyds brigade and was released due to sickness. Says he recovered and spent 18 days on Cotton Hill with militia. Says he was going back to Floyd when arrested. Give oath & release.

Kincaid, James, Jr.—Age 16, lives in Fayette County, son of James Kincaid. Arrested last August, held since. Give oath & release.

Lawrence, P. Dr.—Lives in Fayette Co. Says he was arrested going to join the Virginia militia. No proof against him. Oath & release.

McClung, Alexander—Lives in Nicholas Co. Was in Wise legion and arrested by Floyds men without charges. Give oath & release.

McClung, M.A.—Lives in Nicholas Co. Brother of Alexander McClung. Arrested by General Floyds men the day before the battle of Cross Lanes. Says he is a secessionist and was sent down with prisoners from the Cross Lanes battle. Says he traded the Yankees 18 pounds of butter for nine pounds of coffee. Give oath & release.

Neff, Addison—Age 21, Lives in Fayette Co. No cause for arrest given. He was on a trip from Dogwood Gap to Greenbrier Co. to see his wounded brother. Stopped for a pass at Meadow Bluff and was arrested. No evidence against him. Give oath & release.

O'Dell, Felix S.—Age 26, Lives in Nicholas Co. Says he was at General Floyds camp, taking clothes to his father, when arrested. Claims entire loyalty to the south, has taken oath.

O'Dell, John W.—Lives in Nicholas County. Brother of Felix O'Dell arrested on suspicion only. Give oath & release.

Rader, Anthony—Lives in Nicholas Co. Arrested on suspicion only. Give oath & release.

Ramsay, Samuel—Lives in Nicholas Co. Arrested on suspicion only. He is a Union man but professes loyalty to the Confederate Government. Give oath & release.

Scarborough, Isaac—Age 51, Lives in Fayette Co. Says he was near his home when arrested. Says he was taking a load of beeswax and ginseng down to Kanawha to sell. Arrested by Caskies rangers, they took his horse. Has had long imprisonment. Give oath & release.

Short, Samuel—Lives in Fayette Co. Was arrested by independent scouts. They took two horses from him which were not returned. This man is a known secessionist. Give oath & release.

Siers, Isaac—Lives in Nicholas Co. Says he was in Col. Tompkins regiment and was wounded in a skirmish near Charleston VA. Evidence suggest he is a deserter. Should be turned over to military dept.

Stone, Henry—Age 19, Says he belonged to the 2nd Ky. Regiment. Joined it at Gauley Bridge. Was sent by General Cox to find out where the Confederate militia were in Fayette Co. Stayed his first night at Huddlestons on Cotton Hill. Should be held as a spy.

Stover, F.—Age 16, Born in Raleigh Co. Has two brothers in Cpt. Adams company, Floyds Brigade. Has never seen the Yankees. No proof of anything against him. Give oath & release.

Stover, Sampson—Lives in Raleigh Co. Says he is a southern man. No evidence against him. Give oath & release.

Wardup, William—Lives in Greenbrier Co. Says he was arrested because he expressed the opinion that the Confederate forces would be driven out of the Kanawha Valley (which they were).

White, Robert—Lives in Fayette Co. A feeble old man of seventy. Proves to be a man of good character. Give oath & release.

Williams, Alexander—Lives in Nicholas Co. Arrested on suspicion. Says he did not vote on the secession question. Give oath & release.

Williams, Isaac—Age 51, Lives in Fayette Co. Arrested by Caskies Rangers on suspicion. Man of good character. Give oath & release.

Wills, J.—Age 30, Lives in Raleigh Co. Thinks he was arrested from a malicious charge by Jasper Cole that he is a Union man. Says he is not a Union man. No evidence on him. Give oath & release.

Wills, William—Lives in Raleigh Co. Does not know why he was arrested. No evidence against this man. Give oath & release.

Wriston, Caleb—Lives in Fayette Co. on Johnson Branch of Loop Creek. Says he gave supplies to Caskies Rangers and Jenkins cavalry. Says the Yankees threatened the men of his Branch because they were secesh. Give oath & release.

Wriston, John—Brother of Caleb Wriston. Says he was arrested by Caskies Rangers while at the mill. Says he is a strong southern man. These two men were probably arrested by mistake. Oath & release.

TOTAL: 51

The average age of these prisoners was thirty-three, the charge was generally suspicion of being in sympathy with the enemy. The youngest prisoner was sixteen, the oldest, seventy.[7]

Bushwackers attacked Union pickets of the 30th Ohio Infantry posted at McCoys Mill (present-day Glen Jean) on April 10, 1862. Albert George, a member of the 30th Ohio described the attack in a letter to his wife: "At McCoy's some bushwackers fired on the pickets after night. 48 buckshot and two slugs struck their shanty (made of bark) but as luck would have it missed them, quite a narrow escape. The pickets both fired their rifles and struck the tree that they were behind."[8]

When not skirmishing or fortifying many of the Yankee troops would wander through the Fayette County wilderness in their free time and explore the endless range of hills and valleys. One such sojourn was described by William Ludwig, a member of the 34th Ohio Infantry posted at Fayetteville:

> Camp Union, Fayetteville, Va. April 29, 1862
> Brother George,
>
> I received your welcome letter last night from Mr. Warwick and was very glad to hear from you, and that the folks were all well and that Ben was getting along so well. I am still enjoying good health. The weather is very rainy, yesterday was the first day we have had for nearly two weeks, it rained seven days, until yesterday when we had a fine day and they nearly drilled us to death. Double-quick for two hours at a time, but today I think we will not have the chance to drill, as it is raining again as usual, and nobody knows how long it is going to rain, for when it begins in this weather nobody knows when it will stop. We did have a fine day last Sunday, and I had a fine time, John Kirk and I took a walk out into the country up a place called Laurel Creek. We saw some of the prettyest country in Virginia, the most romantic, the road on both sides lined with fine Laurel, from which I expected every moment to see some bush wacks blaze away from behind the bush and we had no arms but a revolver of Kirks but we each had a good pair of legs which are a very handy thing in a tight place, but we met nothing of the sort, but we did run against a good dinner and a good Union man, we stopped in at the house and he asked us to dinner, of course we did not refuse for a chance to play zouave on ham and eggs, we met them out in the open field and did justice to the 34th Regiment. It was the only thing that a white man could eat I have seen since I left home. I would run the risk of being shot for such a dinner as that. We had a long talk with the man and found him a good Union man, but such as him are very scarce here. In coming home we lost our way some four miles but arrived in camp just before dark. Tell the folks I am well.
>
> Your Brother
> Will Ludwig[9]

Early in May General John C. Fremont, who had recently assumed command of the Union's Mountain Department, decided on a raid against the Virginia and Tennessee Railroad in southwest Virginia. The movement of large numbers of troops through Fayette County on their

Captain W. D. Thurmond. He organized a company of Confederate Partisan Rangers from Fayette and nearby counties.
COURTESY DR. OTIS K. RICE, HUGHESTON, WV

way south was the first significant campaign in this area during 1862. Many of the Yankee soldiers passing through Fayette County at this time had served here previously and this was a sort of homecoming for them. The return of the 44th Ohio Volunteer Infantry to Gauley Bridge on May 10, 1862, was described by Sergeant Bill Lyle:

> The regiment encamped at Gauley Bridge, within a short distance of the camp ground occupied about five months before, and under the very shadow of Cotton Mountain, whose rugged summit the last lingering rays of the setting sun were illuminating with the soft, dreamy splendors of an early Summer's eve, as we drove our tent pins and stretched our cumbrous sibleys. (a type of tent) "In for another campaign, in this jumping off place of creation," said one, as he dipped his tin cup into the kettle of steaming coffee. "Who wants to go on a scout to Sewell Mountain?" shouted another. "Hold on, partner," said a third; "We are going to Richmond by way of Greenbrier." 'What'll ye bet," said another, "we dont march straight for Newbern and cut the Conthieveracy in two, and distinguish ourselves generally?" "Very likely," chimed in another as he leisurely cut into a chunk of fat pork, using a piece of hardtack for a plate—"Very likely boys, we'll distinguish ourselves generally, and some of us may get ex-tinguished particularly."[10]

With the temporary withdrawal of the majority of Union troops from this area, the late spring and early summer of 1862 passed without any significant military activity. The Confederate rangers of Captain W. D. Thurmond remained active during the summer months and seemed to act with impunity. Their well practiced "hit and run" tactics created havoc against soldiers and Union civilians alike. In early August Thurmond's Rangers reportedly burnt the home of Laban and Mary Jane Gwinn, who lived at Round Bottom, near McKendree, Fayette County. Though Laban's brother Samuel had joined the Rebel army, Laban preferred to remain with his family and word soon got around that he was a "Union man." Laban Gwinn was thirty-three years old in 1861 and his wife, Mary, was twenty-three. They had two children; Sarah, six, and John Henry, three. The Gwinns lived a somewhat precarious existence during the first eighteen months of the war, as did numerous other families, regardless of their sentiments. For a while it had appeared that Laban would be able to continue working his farm and supporting his family. With the reported acts of Thurmond's Rangers, coupled with those of bushwackers, the Gwinn family became Union refugees from Fayette County. Two military passes issued to Laban during the war and now owned by his descendants, reveal the circumstances of his move.

Camp McCoy's Mill
August 24, 1862

Capt. Levering,
Dear Sir:

    The bearer Laban Gwinn, a good Union man with his family intends to go to Indiana, he is in reduced circumstances, the bushwackers robbed him, you would oblige me by giving him a pass for one of the government boats to reach Ohio.

                                            Very Respectfully,

Camp McCoy's Mill   M. Stumpf Capt.
Fayette County West Va.   Com. Post
August 30, 62

Guards & Pickets
    Pass Laban Gwinn and family through the lines to Indiana. They are Union refugees.

                                    By order of George Boehm
                                           Capt. 37th Regt O. V.
                                                    Comdg Post [11]

    Laban and his family remained in Indiana for the duration of the war, returning to Fayette County in the summer of 1865. For some unexplained reason the name of Laban Gwinn appears on a clothing list of Thurmond's Rangers. The Gwinn descendants have letters and other documents showing Laban was in Indiana with his family during the war and could not possibly have served as a Confederate Partisan Ranger. The following list is a partial roster of members of Thurmond's Rangers.

ALPHABETICAL LIST OF NAMES RECORDED IN NOTEBOOK OF CAPTAIN WILLIAM D. THURMOND, LEADER OF THURMOND'S COMPANY OF PARTISAN RANGERS, CSA, SHOWING CLOTHING ISSUED TO MEMBERS OF THE COMPANY BETWEEN OCTOBER 1862 AND SEPTEMBER 1864.

Acord, Eli	Crawford, James L.	Hill, John H.
Acord, James	Crone, Michael	Hughart, William H.
Acord, John (1st Cpl)	Davis, Peter	Humphrey, Elijah
Acord, Joseph (1st Sgt)	Davis, Samuel	Humphrey, Lewis
Adkins, Harrison	Deen, William	Hunter, George W.
Adkins, James H.	Dempsey, John E.	Hutchison, David
Adkins, Lewis H.	Duncan,Wm.N.(5th Sgt)	Hutchison, John A
Allen, Henley	Dunn, John (4th Cpl)	Jammeson, John W.
Allen, Perry	Ellison,William(3rd Cpl)	Jeffries, Wm. H.
Arbaugh, James O.	Fink, Harvey M.	Johnson, William H.
Arbaugh, Michael P.	Fisher, Harrison	Johnson, Zachariah
Argrave, James	Fisher, Isaac	Jonson, Thomas

Ayres, James H.
Beard, Thomas L.
Bibb, W.L. (1st Lt.)
Birchfield, Abner
Blake, George
Blake, William J.
Bowles, Irvin
Bowyer, Van B.
Bragg, F. E.
Bragg, Hazard
Burdett, Giles M.
Burdett, Joseph G.
Canada, Jesse B.
Coke, George W.B.
Coleman, Charles M.
Coleman, Seaton
Cook, M. E.
Craft, William J.

Forsythe, Abram
Foster, Andrew A.
Fox, James P.
Fox, Perry
George, Thomas A.
Gill, George
Grimmit, lewis
Gwinn, A. H.
Gwinn, Laban
Gwinn, Marion
Hamilton, William E.
Hedrick, David
Helvey, Boltzer
Hendrick, T. P.
Heslep, Joseph L.
Heslep, Samuel T.
Hicks, James M.
Hicks, Michael

Killy, John
Kincain, Harrison
Kincaid, Sampson
Kishner, J. N.
Leach, B. T.
Lewis, William G.
Lusher, George
Marion, J. Francis
Marrs, James
McComas, Burwell
McCoy, Philip
McNeer, John C.
McNeil, John
Meadows, Richard
Meadows, Valentine
Miles, John T.
Miller, John A.
Miller, William E.

CLOTHING ISSUE LIST, THURMOND'S PARTISAN RANGERS
COMPANY (CSA)—10/1862, 9/1864 continued...

Moody, George
Moody, John
Noble, Edward T.
O'Neel, John
Painter, James K.
Painter, lewis H.
Pegrum, Calvin W.
Peoples, George W.
Poteet, John
Richman, John
Ripley, William R.
Rodes, Greene (2nd Sgt)
Rodes, James Y.
Rodes, Joel Y.
Rodes, John
Ross, John T. (3rd Lt)
Sanger, Joseph (2nd Cpl)
Shepard, James C.
Shepard, William M.
Short, Bartlett
Shumate, John N.
Sidenstricker, George
Smith, Lewis J.
Smith, Thomas
Stinnett, William
Summerfield, Benjamin

Surber, William H.
Taylor, H. Thomas
Tesley, Samuel
Thurmond, Philip J. (Capt)
Thurmond, R. C. (2nd Lieut)
Thurmond, Robert G. (3rd Sgt)
Thurmond, William D. (Capt)
Tincher, A. M.
Toney, Jesse
Toney, Robert
Turley, Samuel
Vandal, Joseph D.
Wade, John H.
Walker, John W.
Ward, Robert
Warden, David J.
Warren, Mathew M.
Warren, William W.
Wees, George
Williams, Daniel
Williams, Floyd
Williams, Linus
Williams, Robert
Windsor, Anderson
Windsor, Isaac
Wood, Felix A. [12]

TOTAL: 142

Laban Gwinn 1828-1900.
Union refugee from
Fayette County.
COURTESY LEONA G. BROWN,
ARBOVALE, WV

Mary Jane Gwinn
1838-1910.
COURTESY LEONA G. BROWN,
ARBOVALE, WV

Colonel Joseph A. J. Lightburn, commander of the Union troops who were driven from the Kanawha Valley in September 1862.
COURTESY WVU ARCHIVES

■ Chapter Eight
Battle of Fayetteville

The lack of active campaigning in Fayette County during the spring and summer of 1862 gave many of the Federal troops time to reflect on their situation and enjoy the relative ease of garrison duty. George W. Botkin of the 1st Regiment, Kentucky Infantry (US) posted at Gauley Bridge, wrote to his wife on August 7:

> This is a pleasant morning. Everything looks lovely. The hills seem to smile as they rise in granduer before me. All nature seems rejoicing, and I feel like rejoicing too. For I am in excellent health, and our hard times seem for the present to be over. Our boys are all well, and in fine spirits, and seem to be as happy as coons in a corn field. We have a pleasant and healthy place for our camp. A high hill rises in our rear, covered with an orchard which supplies us with apples. About thirty yards in our front is the great road leading from Charleston to Richmond. Beyond the road a few paces rushes by the waters of the Great Kanawha, inviting the soldiers to wash their clothes, bathe & etc. Beyond the river rises a hill several hundred feet high, covered with trees, that seem to form themselves into waves, as one gazes on them from the hill above camp. Up and down the road for two miles are seen wagons and tents innumerable. We have plenty of the best water that "OLD VIRGINIA" can furnish; which is the richest luxury that the weary soldier can enjoy. We have seen pure crystal waters gushing forth from the sides of these hills, bidding the soldier a hearty welcome to its refreshing draughts. I have often dreamed of these scenes when reading some exciting novel, but now what used to be a dream is to me a grand reality. I have had some hard times since I first invaded the sacred soils of Virginia, but what I have seen in the Kanawha Valley has well paid me for all my toils and privations. I expect we will remain here for several weeks and I assure you I do not want to stop in a grander or more healthy place.[1]

On August 11 the Federal Government directed that five thousand men be retained in the Kanawha district and that the remainder be brought to Washington, to be used in the more active eastern theater of war. This movement began on August 14. During the ensuing days light draft steamers transported 5,000 men, 1,100 horses, and 270 wagons from the Kanawha Valley.[2] On August 17 General Cox transferred command of the Kanawha district to Colonel Joseph A. J. Lightburn with headquarters at Gauley Bridge. The troops composing Lightburn's command consisted of: The 34th and 37th Ohio Infantry, with four mountain howitzers and two smooth-bore field pieces, under command of Colonel Edward Siber, at Raleigh Court House, with two companies of infantry as a guard for trains at Fayetteville; the 44th and 47th Ohio Infantry, with two companies of Virginia cavalry, at Camp Ewing, ten miles east of Gauley Bridge, under command of Colonel Samuel A. Gilbert; two companies of the 9th Virginia Infantry, and two companies of the 2nd Virginia cavalry, under command of Major Curtis, at Summersville; the remainder of the 9th and 4th Virginia Infantry and 2nd Virginia Cavalry were stationed at different points from Gauley Bridge to Charleston, including an outpost at Coal River in Boone Co.[3]

This reduction of Federal strength in the Kanawha Valley did not pass unnoticed by Confederate forces. As early as August 18 Confederate General William W. Loring began planning an attack into the region. Loring sent his cavalry on an extensive sweep through the area north of the Kanawha Valley. Brigadier General Albert Gallatin Jenkins led the raiding party, starting from the Salt Sulphur Springs in Monroe County on August 22 with 550 men.[4]

When news of the raid reached Colonel Lightburn he became concerned for the safety of his flanks and rear, which were unprotected. He ordered Colonel Siber to fall back from Raleigh Court House to Fayetteville, and Colonel Gilbert also to fall back from Camp Ewing on New River to Gauley Mountain, or Tompkins farm. Fearing that Jenkins would attack Summersville, he ordered Colonel Gilbert to send six companies of the 47th Ohio Infantry to reenforce that point. He also directed a large portion of the quartermaster and commissary stores to be shipped to Charleston.[5]

Jenkins's daring raid proved highly successful. His men skirmished with Federal forces in Randolph, Upshur, Lewis, Roane and Jackson counties, crossing the Ohio River on September 4 and becoming the first Southern troops to raise the Confederate flag on Ohio soil. After a brief skirmish at Point Pleasant on September 5, Jenkins reassembled his force and moved to Buffalo, in Putnam County.[6] With Jenkins camped in rear of Colonel Lightburn's army, General Loring began marching toward the Kanawha Valley. Loring's army of approximately 5,000 men left the vicinity of Giles Court House on September 6.

The news of Jenkins's raid and the anticipated movement by Loring caused a great deal of confusion and excitement among the Federal forces. Samuel J. Harrison, a member of the 44th Ohio Infantry posted at Gauley Mountain, made the following entry in his diary on September 5:

> The Regiment was aroused up this morning about 3 oclock in a great hurry, & ordered to load our pieces, & the Battalion was formed in line of battle in less than five minutes & our Major marched us out on the hill, artillery jackass battery & all as we supposed for a fight from the hurry and excitement although it was all done with almost perfect silence. The teams were also hitched up & ready to move should it be needed. After halting a short time on the hill by camp we were marched back and Col-

onel Gilbert told us there was no sight for a fight he guessed tonight. So the Major marched us in camp and we went to our quarters & stacked arms by companies & had orders to be ready to fall in at a minutes notice as we were likely to be called up again before morning but there was no more alarm & we got to rest till morning. In the morning there was a detail of 150 men 20 corporals 2 or 3 Sergeants 2 Lieutenants & a Capt. we were marched out some three miles on the Lewisburg pike to cut timber, blockade & clear away the trees to give us a fair chance at the rebels if they came at us. We worked at it till noon or past & then was relieved by a new squad & we then came to camp and got dinner & rested the balance of the day. There is various reports about the rebels going to attack us. Some say our supplies are cut off & the telegraph wire cut by the rebels, but I doubt it.[7]

View of Tompkins farm looking west.
COURTESY RUTHERFORD B. HAYES LIBRARY, FREMONT, OHIO

View of Fayetteville drawn September 10, 1862. The town was attacked by Confederate forces later that same day. COURTESY FAYETTE COUNTY HISTORICAL SOCIETY

On September 10 the anticipated attack against Fayetteville occurred with the advance of Loring's army. Arriving four and one-half miles south of town at 11:15 a.m., the 51st and 22nd Virginia Infantry with Clarke's Battalion of sharpshooters were sent by a road to the left in order to attack the Yankees in rear, while the main body proceeded directly along the turnpike to attack in front. The Confederates were guided on this flank movement by Benjamin Jones, father of Buehring H. Jones, Colonel of the 60th Virginia Infantry. As the main body advanced to within two miles of Fayetteville the 26th Battalion Virginia Infantry was attacked by three companies of Federal infantry hiding in the woods on both sides of the turnpike. As a sharp and stout resistance was offered by the Yankees, Brigadier General John S. Williams, commanding the Confederate 2nd Brigade, deployed a large force to the left and right of the enemy. After a short skirmish, hotly contested, the Yankees withdrew towards Fayetteville.

As Loring's troops continued their advance they were again attacked by Federal skirmishers one-half mile south of town. In this attack the Yankees were hiding in a dense laurel thicket alongside the road and poured a deadly fire of minnie-balls into the Rebel column. Two artillery pieces were ordered brought up and the skirmishers were quickly cleared. From this position there were three small hills between the Rebels and the first Yankee fort. The 45th Virginia Infantry was sent up the side of the first hill, which was in easy range of the enemy guns, and used to divert the enemy's attention while a portion of that regiment moved by a flank movement through the woods to the next hill. In this position the Confederates remained quietly until the troops holding the first hill were brought up and placed to their right. When the entire regiment was in place they opened a heavy fire on the Yankees, driving them from a house near their fortifications. Firing of the Federal troops was also heavy, and it was reported by a Confederate that "the grape-shot and minnie balls flew thick as hail around us."[8]

After considerable brisk fighting in this position, and when the Federals had been driven to their stronghold, the Rebels advanced obliquely to the left to a position in the woods within 100 yards of the enemy fortifications. While severe fighting was taking place in front, the forces sent on a wide flanking movement to the Federal left reached their destination at 2:15 p.m. and found that the enemy batteries were not in the position which had been described. They found in front two batteries well constructed, and so arranged as to command, by a cross-fire, the cleared space of about 1,000 yards between the batteries and the wooded ridge on which the Rebels took position, Through this cleared space ran the turnpike from Fayetteville to Gauley Bridge. Due to the excessive heat of the day, and their long fatiguing march by a circuitous route, the Rebels were nearly exhausted upon reaching their destination and had yet to attack the enemy.

Under these circumstances, after consultation among some of the officers, the Confederates decided to take and hold a position commanding the turnpike leading from Fayetteville to the Kanawha River, to prevent the passing of enemy trains, and, if possible, cut off their retreat. To accomplish this, three companies of the 51st Regiment were directed to take position on a spur extending out and commanding the turnpike on the Rebel extreme left, and about half a mile in rear of the enemy batteries. Sharpshooters were instructed to take position on the immediate right of the 51st Regiment to prevent them from being flanked. Colonel George Patton, with a portion of the 22nd Virginia Infantry, was thrown farther to the right to occupy another spur, commanding, with long range guns, another section of the turnpike. Major Robert A. Bailey, also of the 22nd, was sent with a detachment to the extreme right and nearer the batteries; the remainder of the forces, as they came up, were held in reserve to support any part of the line that might be attacked, and also to charge the Federal artillery if an opportunity should arise.[9] Before the Confederates had successfully completed their deployments they came under fierce attack by six companies of infantry ordered out by Colonel Siber. Three times the Yankees charged the knoll held by the 51st Virginia Infantry, causing the Rebel ammunition to become so scarce that the men were told to fire only as the Yankees advanced up the hill. Three hours of "murderous and unequaled" combat inflicted serious losses on men and officers of the Federal units.[10]

As the battle in front raged on, drummers of the 37th Ohio Infantry carried water to their weary friends. John S. Kountz, a member of the 37th Ohio, described the events he witnessed: "During the engagement the drummers carried water to the men from a well on the Fayetteville road, an exceedingly hazardous employment, as we were obliged to pass an open space exposed to the enemys fire. The 34th on our right fought gallantly in an open field, and charged the Rebels several times, sustaining heavy loss, one-half of the officers and fully one-third of the men engaged being either killed or wounded."[11]

The Confederate artillery initially took position on an eminence 500 yards from the first enemy fort. When they advanced to within 300 yards they came under a scathing fire from Federal artillery and sharpshooters, who held the ravine and hill opposite them. A Confederate officer described their situation as a "storm of grape and canister."[12] The attack in front could not be fully supported by infantry because the Yankees were so strongly entrenched their position could not have been taken without great loss of life. The frontal attack was more on the order of an artillery duel, as described by Fountain G. Shackelford, a member of the 36th Virginia Infantry:

153

Colonel Eliakim P. Scammon, commander of Federal forces at Fayetteville and vicinity 1863.
COURTESY MASSACHUSETTS COMMANDERY MILITARY ORDER OF THE LOYAL LEGION AND U.S. ARMY MILITARY HISTORY INSTITUTE

The 36th supported Bryans Battery which did good work, although they lost some men in killed and wounded, and the guns injured somewhat. While this was going on, our regiment was lying behind the hill upon which the artillery was placed in the hot broiling sun, without water or any to be had conveniently. It seems to be very strange but true, that men are prone to sleep while an artillery fight is going on if not engaged. During this fight there were numbers of men that slept soundly in the broiling sun, while the cannons booming and the shells flying over their heads. A citizen would never have slept on account of excitement, but not so with a soldier. When a soldier gets into a fight he loses all fear as to what the consequences will be, and takes everything calmly and quietly as if nothing was going on.[13]

As the battle raged, Colonel Lightburn, who was then at Gauley Bridge, ordered three companies of the 4th West Virginia Infantry to advance into Fayetteville, and five companies of the 47th Ohio Infantry to take position on top of Cotton Hill, to be of assistance in the event of a retreat.[14] Lightburn's troops posted in the vicinity of Gauley Bridge were ordered to pack everything and to destroy what they could not quickly save. Samuel Harrison of the 44th Ohio, posted at Tompkins farm, described the scene in his diary:

> Directly after dinner, about 12 oclock, we heard the firing of artillery in the direction of Fayetteville which soon became frequent & distinct & the smoke which soon became visible showed that the 34th & 37th had been attacked. We worked on until about 3 p.m. when we were ordered to camp in considerable hurry. When we got there the tents were struck and all hands busily engaged in packing & were about ready to move. A good deal of excitement also prevailed. We also learnt that they had been fighting all afternoon & were getting over powered. The Battalion was formed about six oclock and we were then ordered to make a little coffee & just after dark we were marched out towards Gauley taking all we could & burning what commissary stores we could not haul. About 7 or 8 oclock we reached Gauley Bridge where we lay over till morning waiting for orders.[15]

Just after sunset, as the attack on the Union rear had ceased and the frontal attack slowed somewhat, the men of the 37th Ohio Infantry were ordered to fix bayonets and charge a portion of the Rebels in front. This bayonet charge, the only one ever used in Fayette County, succeeded in driving some of the Rebels back, and even pursued them a short distance along the turnpike. As the charge developed the reenforcements sent by Colonel Lightburn, along with four companies of the 34th Ohio, and twenty-five horsemen of the 2nd Virginia Cavalry, arrived.[16] Lieutenant Colonel John L. Vance, of the 4th West Virginia Infantry, described his arrival at Fayetteville in a speech he gave in 1896:

> The men were in good condition and moved rapidly. Before the top of Cotton Hill was reached, we met a number of civilians who gave infor-

mation that Fayetteville was surrounded; that Siber's force was cut to pieces; that when they last saw the town, it was burning, and in truth, all the stories which would naturally be told at such a time and under such circumstances. As we continued to advance, the number of refugees, seeking shelter at Gauley, increased, and the stories grew in horror. When near to Fayetteville, we saw evidences of conflict and in the woods to the right of the road, within short rifle range, several bodies of troops were plainly seen, supposed to belong to the enemy and such afterward proved to be the case. It was after sundown but not dark, when my command entered Fayetteville. I reported at once to Colonel Siber. He received me cordially, thanking me for coming to his aid, asked the number of men and a few other questions of like character. In response to his inquiry, I said we had not been fired upon. He expressed surprise, saying "We are surrounded." After instructions in regard to the disposition of my command, he gave an outline of the proceedings of the day.[17]

By 9:00 p.m. all fighting had ceased with Union and Confederate forces maintaining virtually the same positions they had held since late afternoon. The fact that one Federal regiment and six companies of another with four mountain howitzers and two six-pounder smoothbores, had held at bay five thousand Confederates with 16 pieces of artillery, was a brilliant feat for which Colonel Siber deserves the highest praise. Confederate casualties at this time consisted of approximately 17 killed and 32 wounded. Federal losses were approximately 13 killed and 80 wounded.

At 10:00 p.m. Captain R. L. Poor, chief engineer for Loring's army, broke ground for an artillery position a scant 130 yards from the first Yankee fort. Working in the darkness, Poor's crew quickly completed a position suitable for two siege guns, a 24-pounder howitzer and 12-pounder rifle gun.[18] The Confederates did not have an opportunity to spring their surprise on the enemy. During the late evening hours Colonel Siber began the laborious process of transporting his wounded and wagon train out of the area, and at 1:00 a.m. he ordered a retreat. A quantity of commissary stores were set fire and the retreat was effected without any molestation but from a company of Rebels hidden in the woods to Siber's rear. During this late night skirmish the Rebels captured a small section of Siber's train and wounded several of his men. For some unexplained reason the Confederates failed to make a serious attempt to halt the Federal retreat and Siber's men reached Cotton Hill at dawn where they joined five companies of the 47th Ohio sent to that position earlier by Colonel Lightburn.[19]

Shortly after Siber's troops reached the top of Cotton Hill they observed Loring's entire army with colors flying in order of battle deploying at the base of the mountain. The Yankees had no time to rest or regroup; quick action was necessary if they were to avoid being cap-

General William Wing Loring. He captured Fayetteville and Charleston, WV in 1862 but retreated after a few weeks in the Kanawha Valley. His decision cost him his command. COURTESY SWV

Colonel George S. Patton, 22nd Regiment Virginia Infantry.
AUTHOR'S COLLECTION

tured or killed. Colonel Siber ordered the main body of his troops to continue retreating, while five companies of the 37th Ohio Infantry with a detachment of artillery remained behind to slow Loring's advance. Loring divided his force, sending half along the old Fayette road and the other half directly up Cotton Hill along the new turnpike. Acting swiftly, Siber's men opened on the Rebels with artillery and musket fire from behind a hastily prepared breastwork. A general engagement ensued with both sides firing large quantities of ammunition in a short period of time. The Confederates attempted to advance as they fired, but had the disadvantage of position making progress difficult and hazardous. Siber's resistance was so strong that the Rebels were driven off Cotton Hill at 10:00 a.m., with a loss of two killed and three wounded.[20]

With Loring's army regrouping at the base of the mountain, Colonel Siber seized the opportunity to make good his escape. A rapid, disorderly march was resumed, with the Federal situation growing worse as the entire scene became more of a rout than an organized withdrawal. As was generally the case with any army during a rapid flight, Siber's men threw away guns, knapsacks, camp gear, anything that would slow their progress. The Confederates resumed their pursuit with a running skirmish taking place all the way across Cotton Hill and down the western slope near Montgomerys Ferry, below Kanawha Falls. Fountain G. Shackelford, a member of the 36th Virginia Infantry, described what he witnessed as he descended Cotton Hill:

> We hastened on and soon arrived at the falls. The regiment was but little ahead of us. We there beheld the greatest destruction of army stores, that we saw during the war. The whole valley from the falls to the bridge, a distance of two miles, was enveloped in smoke. They aimed to destroy everything, that nothing might fall into the hands of the Confederates. Although we got nothing, numbers of the citizens visited the place, and found hundreds of dollars worth of property that had not been destroyed, such as wagons, iron horse shoes, salt, bacon, and many other things, which was of use to the citizens; all of which they appropriated to the necessities. It was said at that time the army stores destroyed at that place, amounted to two million dollars; whether it was that amount or less I do not know, but I know it must have been immense, because it was the depot of supplies for all this section of the country where the Federal army operated.[21]

Another fight took place at Montgomerys Ferry, as members of the 44th Ohio Infantry took position on the river bank to slow Loring's advance. Samuel J. Harrison of the 44th Ohio described the fight in his diary:

> Thursday, September 11, 1862—Weather dry and very hot and the roads extremely dusty. We got breakfast and stayed in Gauley till 7 or 8 oclock a.m. burning & destroying some commissary stores & then marched down to the ferry about one mile below the bridge, & formed our line of

battle on the bank to cover the retreat of the 34th & 37th which was now being made as fast as possible they were over powered by a superior number of the rebels losing something like 100 men in killed and wounded. The officers of the 34th suffering heavy from the fire of sharpshooters. About 11:00 a.m. a large force of the rebels made their appearance on Tompkins farm where we had left the evening before. The 34th & 37th train was ferried over safe & the two Regiments passed down on the other side of the river when the rebels made their appearance in pursuit of them. Our company & company K being left to support the artillery, as soon as the rebels showed themselves we opened a heavy fire on them which caused them immediately to fall back and take shelter in the timber when a general bushwacking & cannonading began which lasted sharply for about an hour & a half, when we were ordered to fall back on the reserve in double quick. Our loss was one man (Lafayette Rafe) wounded. Although the bullets fell like hail among us. Before the firing began however, we burnt Gauley Bridge, a vast amount of commissary stores, blowed up the magazine & burnt the ferry boat. The 47th was also cut off in falling back from Somerville & had to burn their train & take the mountain paths to keep from being captured. We retreated back till night through a suffocating dust & heat, the rebels following up close & artillery fighting in the rear occassionally all evening. We bivwaced near Candleton about 1/2 the night for a little rest & sleep.[22]

As the battle at Montgomerys Ferry continued, Loring ordered Lieutenant Joel H. Abbott to pick a squad of men and take a cannon along the ridges of Cotton Hill to the cliffs overlooking Gauley Bridge. This assignment was cheerfully met by Lieutenant Abbott and in a short while Confederate artillery was once again blazing away at Gauley Bridge, just as it had done the previous fall under command of General Floyd. It was from this bombardment of Gauley Bridge that the famous Cotton Hill cannon story originated. Writing after the war, Lieutenant, later Captain Abbott, claimed that he and his men abandoned the cannon when they found it too difficult to remove because of the exceedingly rough terrain. Abbott wrote: "...trying to get the gun back and finding it a difficult job, we hid it in a deep ravine, and it is there yet."[23]

Whether or not a cannon was actually abandoned on Cotton Hill, and if so, whether or not it is there yet, is purely a matter of conjecture. Captain Abbott was the only member of the squad who had the gun, to later claim it was left on the mountain. One thing is for sure, the Cotton Hill cannon story has, over the years, taken on the aura of a legend. Numerous people from various parts of the country have braved the rocky trails and deep ravines of Cotton Hill in search of the "Captain's Gun." In my travels throughout West Virginia and Virginia I have often been asked if I know anything about the Confederate cannon at Cotton Hill. At times it has even been described as the "gold cannon." Unfortunately, I do not know much more than others interested in the story. I do know that Captain Abbott did this area a favor, intentionally or not,

Captain Joel H. Abbot of the 8th Virginia Cavalry. He claimed to have abandoned a Confederate cannon on Cotton Hill in 1862.
COURTESY GEORGE ABBOT FISHER, BOOMER, WV

as his story has put Fayette County "on the map."

As the fighting along the river bank slowed, Confederate General John S. Williams, called for volunteers to swim across the Kanawha and retrieve a ferry boat the Yankees had tied to shore and set on fire. In answer to his call, Dr. Watkins, of the 36th Virginia; Lieutenant Samuels, of Williams's staff; W. H. Harmon, and Allen Thompson, of the 45th Virginia, and several others, jumped into the river and boldly swam, under a shower of grape and canister. They seized the burning boat, made fire buckets out of their hats, extinguished the flames, captured a Yankee officer and ten men, and brought the boat safely to shore. Williams then ordered Colonel McCausland, with the 36th and 22nd Virginia Infantry, and two pieces of artillery, to cross the river and pursue the enemy. He then used the balance of his force to chase the Yankees along the left bank, skirmishing almost constantly.[24]

The running pursuit continued into the night of September 11 and within a few hours there were no Yankee soldiers to be found in Fayette County for the first time since the previous summer. Destroyed property marked the route of the Federal retreat, offering Loring's men a rare opportunity to supply themselves with new guns, clothing, camp equipage and food. A Rebel commissary agent noted that during the offensive the Confederates had captured 20,000 pounds of bacon, 425 wagons, 300 bags of oats, and numerous quartermaster stores.[25]

There was little fighting on September 12 as the pursuit neared Charleston. On September 13 the Federals formed a line of defense near the Elk River bridge, and warned the citizens of the impending attack. Many of the civilians were sent to Cox's hill on the north side of town and told to remain there until the battle was over. The Yankees then set fire to their supplies, which were stored in large warehouses between Virginia and Quarrier streets and Capitol and Hale streets. Loring's troops soon made their appearance, which resulted in a hard-fought battle for the possession of Charleston. Once again Lightburn's troops were driven from their position after several hours of artillery and musket fire. As the Yankees fled Charleston they destroyed the Elk River bridge and struggled to retain possession of their wagon train which was said to be 700 or 800 wagons strong, covering a distance of seven miles.[26]

With the retreat of Lightburn's army came an exodus of Union citizens and others who feared reprisal by Confederate commanders. Coalsmouth resident Victoria Hansford Teays, who had a brother in the Southern army, recorded the exodus in her journal. She wrote:

> Such a sight about me I never saw nor ever expect to see again. The river as far as the eye could reach up and down was covered with boats of all kinds, large flat boats, jerry boats, jolly boats, skiffs and canoes. In these were all kinds of people and all kinds of things.... When I say the river was covered with boats I mean just what I say, and a person could just

almost have crossed the river by jumping from one boat to another. They had been looking for the retreat and had boats prepared. Not the soldiers I mean but citizens who wished to retreat with the Union army. Or those who had acted so shabbily toward the Southern people whose friends and relatives were off with the rebs that they were afraid to stay when the country fell into the rebel hands. Several families were on gunwayles lashed together, and I will never forget a woman and a man who were on two short pieces of gunwayles lashed together. A tub sat at one end of it containing their property. The woman sat in a rocking chair at the other end, while the man stood in the middle and paddled them on as best he could. The woman was wet to her waist from the water washing over the planks, the man seemed to be wet all over from push-pushing and pulling. I felt so sorry for them and said, "poor fools, the rebels wont hurt you." But on they went. . . . By evening the rebel cavalry came into our village, and the first that came brought the sad news of the death of Jimmie Rust, son of Samuel Rust, who lived exactly opposite the mouth of Coal (River). He was killed at the battle of Fayetteville, killed instantly in the heart of the battle, in perfect health and perfect spirits, thinking perhaps of the home almost in sight, thinking of the mother who was watching and waiting for him, of brothers and sister and father who would welcome him as a brave soldier. He was brought home in his coffin and buried on the high hill over Kanawha at the Rust graveyard. . . . the next morning at scarce daylight my brother came home. I heard an exclamation from my father and ran down stairs to find my brother, Carroll, oh how thankful to have him home again, well, but sunburned and rough enough, after soldiering in his butternut suit. He had been gone over a year, nearly 15 months, and had endured every hardship and privation, as had all or most of the others. Of course there was nothing too good for them. Oh, how we loved and petted them, for we didnt know but that they might never come again. I remember I shook hands with Christopher Crouch on this very day. The 10th of next September he was killed at the battle of Dry Creek (White Sulphur Springs).[27]

The following day, September 14, General Loring issued a proclamation to the people of western Virginia, and a congratulatory order to his command:

To the people of Western Virginia:

The army of the Confederate States has come among you to expel the enemy, to rescue the people from the despotism of the counterfeit State government imposed on you by Northern bayonets, and to restore the country once more to its natural allegiance to the State. We fight for peace and the possession of our own territory. We do not intend to punish those who remain at home as quiet citizens in obedience to the laws of the land, and to all such, clemency and amnesty are declared; but those who persist in adhering to the cause of the public enemy and the pretended State government he has erected at Wheeling will be dealt with as their obstinate treachery deserves. When the liberal policy of the Confederate Government shall be introduced and made known to the people, who have so long experienced the wanton misrule of the invader, the commanding general expects the people heartily to sustain it, not only as a duty but as a deliverance from their task-masters and usurpers. Indeed, he already recognizes in the cordial welcome which the people everywhere give to the army a happy indication of their attachment to their true and lawful Government. Until the proper authorities shall order otherwise, and in the abscence of municipal law and its customary ministers, martial law will be administered by the army and provost-marshals. Private rights and property will be respected, violence will be repressed and order promoted, and all the private property used by the army will be paid for. The commanding general appeals to all good citizens to aid him in these objects, and to all able-bodied men to join his army to defend the sanctities of religion and virtue, home, territory, honor, and law, which are invaded and violated by an unscrupulous enemy, whom an indignant and united people are now about to chastise on his own soil. The Government expects an enthusiastic and immediate response to this call. Your country has been reclaimed for you from the enemy by soldiers, many of whom are from distant parts of the State and the Confederacy, and you will prove unworthy to possess so beautiful and fruitful a land if you do not now rise to retain and defend it. The oaths which the invader imposed upon you are void. They are immoral attempts to restrain you from your duty to your State and Government. They do not exempt you from the obligation to support your Government and to serve in the army, and if such persons are taken as prisoners of war the Confederate Government guarantees to them the humane treatment of the usages of war.

By command of Major-General Loring:

H. Fitzhugh,
Chief of Staff

Loring's announcement to his troops read as follows:

GENERAL ORDERS,   HDQRS. DEPT. OF WESTERN VIRGINIA
No. -   Charleston, W.Va., September 14, 1862.

The commanding general congratulates the army on the brilliant march from the southwest to this place, in one week, and on its successive victories over the enemy at Fayette Court-House, Cotton Hill, and Charleston. It will be memorable in history that, overcoming the mountains and the enemy in one week, you have established the laws and carried the flag of the country to the outer borders of the Confederacy. Instances of gallantry and patriotic devotion are too numerous to be specially designated at this time; but to brigade commanders and their officers and men the commanding general makes grateful acknowledgment for services to which our brilliant success is due. The country will remember and reward you.

By command of Major-General Loring:

H. FITZHUGH,
Chief of Staff.[28]

Confederate losses during the period September 6 - 16, included approximately 24 men killed and 89 wounded. Federal losses were approximately 25 killed, 95 wounded, and 190 missing. Once the Southern forces had established firm control of the Kanawha Valley, General Loring posted at Charleston approximately 4,000 troops, with detachments at Gauley Bridge and Fayetteville. He made no serious effort to extend his operations beyond Charleston, but concerned himself mainly with producing salt at the Kanawha Salines and transporting it to eastern Virginia. Soldiers and civilians alike benefited greatly from these shipments of salt, as described in the diary of Dr. Samuel R. Houston of Monroe County: "Our forces in possession of the Kanawha salt works. Farmers in great numbers going there for the salt in the captured wagons. One million pounds for disposal at 35 cents the bushel. We have been paying $5. The county has been purchasing wheat at $3.50."[29]

Returning to the Kanawha Valley was certainly a glorious homecoming for many of Loring's men who had friends and relatives in the area. Numerous families of Southern sentiment were brought together again for the first time in over a year. Some unfortunate families were greeted by news of the death of loved ones, several of whom had been killed on their march to Charleston. Many of the Yankee troops chased out of the region also had an opportunity to see their loved ones. Captain Tom Taylor of the 47th Ohio Volunteer Infantry was hoping to see his wife and children when he wrote to her while on board the steamboat *Mary Cook*:

> Thursday Sept 18 1862
> Steamer Mary Cook
> My Dear Wife
>
> I embrace this, the first opportunity I have had to give you a few words relative to our condition. I notice by the paper of the 15th that we are completely cut off — Dont believe it. On Wednesday of last week we received orders to move. This order was given near sunset. At 8 pm after burning a vast amount of stores & camp equipage we started on our retreat, or as we suposed to the relief of the forces on the river. On Wednesday they were attacked by a superior force under Loring. The odds were on 4 to 1 but the 37th & 34th held them at bay & compelled them to retire at night. Next morning 4 Co's of the 47th Regt were sent to assist them to cut through & another days hard fighting took place. I had some in that fight who did themselves great credit. We marched hard all Wednesday night & Thursday until noon. Up to this time nothing of importance had occured to us. We crossed over to the 9th Va. and stopped to burn more baggage. At noon we received information that the rebels held Gauley & that were cut off, then we burned *all* our baggage, stores, wagons & everything. We then started a patrol to give us notice of the advance of the enemy while we took dinner, then we studied how to get through. We had not been in the mountains over a year for nothing. We

165

used our knowledge, burned our ambulances, mounted our sick on horses, and dashed single file up a creek. The mercury was at 96 and over a mile we marched through sand over two inches deep. We had many sun-struck, I had one—but we pushed on. Presently we entered a gulch and commenced a toilsome ascent on a very high mountain. We got about 1/2 mile up when our patrol came riding in that a heavy Cavalry force of the rebels were in pursuit. Here we formed our first line of battle in the shape of a "V" point down. Cpt. Wallace on the right and I on the left & waited, on the force came, but just as we were at "ready" I saw a horse I knew and then the rider, Lt. Col. Curtis, Lizzies husband. I ordered—arms, and we greeted our friends who had also been cut off, with three hearty cheers, and started again. After marching through the Mt single file about 18 miles we struck the Charleston road at Cannelton. All our forces except our howitzer battery and the 44th. It was now near 9 oclock pm of Thursday but still the rebels were not more than 500 yards behind and only the angry growl of howitzers kept them back. 1/2 an hour later we would have marched out in the rebel army. Well, "forward march" and the weary boys plodded on, dust two inches deep, little water and many rebels. At 2 am Friday we rested & got breakfast & at four marched, skirmishing almost constantly, four companies of the 47th rear guard on side of the river & six Co's of the 47th advance on the other. We reached Camp Piatt at one, with the rebels so close to our heels that we had to stop, so we constructed a picket and posted our batteries formed a line of battle and waited. While waiting for attack your letter of the 31st August was received. A heavy rain came up and continued during the night and at 12 at night amid the darkness we marched and left the rebels an empty field. Our four Co's crossed & the 47th shook hands at night on 15th—at 8 am we reached Charleston. All the other forces i.e. 34—37 had crossed to get cross the Elk. The 47th was left with two howitzer batteries to fight and hold Charleston. At 9 am our pickets were driven and skirmishing began. At 10 we opened our first piece and the rebels replied from both sides of the river, firing became general and from this time we were under the most murderous fire until 20 minutes after 3 pm when we were compelled to retire because the howitzer ammunition gave out. At 3 the 20 pounder beyond Elk commenced shelling the rebels, then after crossing artillery mingled with the rattle of musketry & was kept up until near 10 pm. We had a narrow escape—but I believe I lost none—I think all will turn up I had none killed or wounded I made my men fire and load lying. Cant you bring the children and come to Point Pleasant to see me? Fix to stay 2 or 3 weeks, I think we will stay there some time. If you come, come quickly, it is above Gallipolis on the Kanawha & Ohio River.

Affectionately
T. Taylor[30]

Almost immediately upon his arrival at Charleston, General Loring began extensive recruitment efforts. Initial success prompted him to write to the Confederate Secretary of War on September 19 requesting

5,000 stand of arms with which to arm new recruits.[31] On September 25 Loring received a letter from General Lee congratulating him on the success of his operations and recommending a foray down the Monongahela Valley through Clarksburg, Fairmont, and Morgantown, to destroy railroad bridges and tunnels, even suggesting that Loring might be able to push his army on into Washington County, Pennsylvania.[32]

Within a few days Confederate success at recruitment fell off sharply and Loring wrote to the Secretary of War on September 28 asking for 5,000 reenforcements and blaming the lower than expected recruitment on the Cincinnati newspapers, which were widely circulated in the area, and on the conduct of Virginia State Line forces under command of General Floyd. Loring claimed that Floyd's men had alienated an otherwise loyal populace through their oppressive conduct, and in addition to that his officers had enticed some of Loring's men into joining the State Line force with the allurement of one years service in contrast with the service of three years in the regular army.[33]

Instead of sending Loring reenforcements, the Confederate Government, on September 30, instructed him to make the necessary arrangements for the defense of the Kanawha Valley, leaving a portion of his army to cooperate with the forces of General Floyd, while he moved into the northwest to break up the insurrectionary government at Wheeling and inflict all possible damage to the Baltimore and Ohio Railroad. In addition, he was to persuade the people that traitors to the South would be punished and warn them that Virginia would never permit a division of the commonwealth. Having accomplished these objectives he was to join forces with General Lee, who was then camped near Winchester, Virginia.[34]

On October 7 Loring wrote to the Confederate Secretary of War, suggesting that the route proposed by General Lee was too long and offering his own plan. Loring stated that it would be advisable for his army to proceed by way of Lewisburg and Monterey, and without awaiting a reply or receiving any instructions, not more than forty-eight hours after he had written, he began to move on the route he had proposed, leaving the Kanawha Valley totally unprotected. His trains were started on October 8, with the cavalry of General Jenkins, some 1,500 strong, remaining behind to delay the Federal advance which he knew was soon to come.[34] Loring's army encamped at Kanawha Falls on October 11, at which time he issued another congratulatory order to his men:

GENERAL ORDERS,   HDQRS. DEPT. OF WESTERN VIRGINIA
No.   Falls of Kanawha, W.Va., October 11, 1862

The commanding general deems it proper to announce to the troops that, having accomplished the object for which they were sent into the country-by driving the enemy, strong in numbers and the insolent sense

General John Echols. He succeeded General Loring in October 1862 as commander of Confederate troops in the Kanawha Valley.

of security, to the borders, and capturing his posts, stores, and many prisoners-in obedience to the orders and the original plans of the Government, this army will prosecute the campaign and move to a new field of activity. He feels confident that on other fields it will earn the commendation it has deserved and received from the country for its late brilliant victories here, and retain its present almost unrivaled reputation for endurance and valor.

That portion of the army from the western frontier who are marched from their homes deserve to know that their services will still be used, when they will avail the most, for the deliverance of the whole of their noble State from the invader and usurper, and that the eye of the Government is turned to the value and welfare of their section and to the merits of men who so cheerfully leave their homes a second time to perform a patriotic duty to their State and to the country. Your unselfish devotion to duty, and the exhibition of loyalty by most of the best people of your section, have fixed more firmly than ever the purpose of the Government with its victorious armies to rescue and retain your country.

By command of Major-General Loring:

H. FITZHUGH,
Chief of Staff.[36]

Loring's decision to leave the Kanawha Valley unprotected and move by a route not approved by the Confederate Government cost him his command. On October 15 he was ordered to report at Richmond, turning his command over to Brigadier General John Echols, who was expected to move back into the Kanawha Valley and reestablish Confederate control. Echols acted swiftly to reoccupy the valley, and within a few days his headquarters were established at Charleston. Echols's stay in the region proved to be shorter than Loring's. On October 27 Echols notified the Richmond government that he had received reliable information that 12,000 enemy troops were within ten miles of Charleston. He also stated that he believed an additional 3,000 enemy soldiers had gone to Guyandotte for the purpose of attacking General Floyd, and that another 4,000 had gone to Clarksburg to endeavor, by marching to Montgomerys Ferry, to intercept his command.[37]

At 2:00 a.m. October 28, Echols began a forced march which continued for a distance of 31 miles, stopping a few miles west of Kanawha Falls. By evening of the next day he had succeeded in passing his entire train over Cotton Hill and had established headquarters at the home of Luther Warner on Laurel Creek near present day Beckwith. This was the same house used as a headquarters by Generals Wise, Floyd, and Rosecrans, the previous year. Echols did not remain long, as he feared the Yankees would now advance into the area rapidly by using steamboats. He reported that the Kanawha River had risen some four feet within the last few days, and it was now possible to come by boat all the way to Loup Creek.[38]

Echols continued his retreat on October 30, and the following day General Jenkins's cavalry skirmished with a superior force of Yankees at Kanawha Falls, which resulted in his being driven out of Fayette County. Echols then ordered him to take his command to the counties of Greenbrier and Pocahontas, and watch and counteract the movements of the enemy from the direction of Clarksburg and Beverly. Echols feared the Yankees would not be content with retaking the Kanawha Valley, but would proceed south and threaten either the Virginia Central or the Virginia and Tennessee Railroad.[39]

General Jacob D. Cox, who had served in the Kanawha Valley during most of 1861, was placed in charge of the Federal forces sent to retake the valley. His army of 9,000 reached Charleston on October 29. The Kanawha Valley was permanently given up to the Yankees, who resumed, under new commanders, the positions they had occupied when the campaign began. Confederate forces never again gained control of the region, and the valuable salt resources of the Kanawha Salines were sorely missed. It had been a brief but joyous return for the Rebels, one they did not repeat until war's end.

■ *Chapter Nine*

Winter Quarters

During their retreat the Confederates burned several small bridges along the route, obstructed the road with fallen timber, and destroyed the flatboats along the river. These acts had the desired effect and slowed considerably the Federal advance eastward from Charleston. By November 3 Gauley Bridge was once again occupied by the Union army. General Eliakim Parker Scammon was placed in charge of the division there and Colonel Lightburn resumed command of his brigade.

Scammon's most pressing task was to reopen roads, make ferries and bridges, and thus renew the means of getting supplies to his troops. The Kanawha River was unusually low for the season and every energy was therefore necessary to get forward supplies to Gauley Bridge and the other up-river posts. General Cox directed Scammon to inspect carefully all of the old Federal positions as far as Raleigh Court House and to report whether the driving of Colonel Siber's troops from Fayetteville had been due to any improper location of the fortifications there. Scammon was also expected to examine the road up Loup Creek, and any others which might be used by the Rebels to flank the posts in the vicinity of Gauley Bridge. With this information Cox hoped to form intelligent plans for a more secure holding of the region.[1]

On November 8 Cox received a telegraph message from the United States Secretary of War, informing him that no posts need be established beyond Gauley Bridge, and that about half of his command should be sent to Tennessee and the Mississippi Valley.[2] It was thus definitively settled that Cox's task for the winter was to restore the condition of affairs in West Virginia which had existed before Loring's invasion. He would also have to organize his district with a view to prompt and easy supply of his posts, and effectively suppress lawlessness and bushwacking. This was a "tall order" indeed, but General Cox set about his duties with characteristic skill and vigor.

With winter rapidly approaching Cox gave priority to the arrangements for transportation. Wagons were scarce and the river was once again looked to for the dependable movement of supplies. General George Crook, who had proven himself dependable while commanding the Yankee post at Summersville, was stationed at Gauley Bridge; Scammon's headquarters were established at Fayetteville. On November 12 reports were received that authentic information showed that Confederate General "Stonewall" Jackson was planning an advance from the Shenandoah Valley into West Virginia. These reports caused quite a stir among Cox's command and everyone was much relieved when they proved false.

Among the Union troops returning to Gauley Bridge was Lieutenant George B. Turner, of the 92nd Ohio Volunteer Infantry. Lieutenant Turner's letters, written from various parts of Fayette County, offer excellent first-hand accounts of the activities of the Federal troops in this region. Turner wrote to his mother from Gauley Bridge on November 15:

> As you see by the heading of this letter we are still at Gauley Bridge.... For several days a large fatigue party detailed from the brigade, have been engaged in drawing from the river the wire ropes that belonged to the fine bridge across the mouth of Gauley. One is surprised on looking at the finely cut stone of the piers still standing that such a bridge should be erected in such a wilderness. It is almost like finding a work of art in a desert.... The boys were really glad to get back into this region, having had enough of the east and eastern service. They do not wish to remain here however and hope to move down the river to winter at some point on the Ohio.[3]

Lieutenant George B. Turner, 92nd Ohio Volunteer Infantry. He served extensively in Fayette County and was mortally wounded at the battle of Mission Ridge Tennessee, November 24, 1863.
AUTHOR'S COLLECTION

The following day, the 16th, Turner wrote to his father: "I have good health, enough to eat, enough to do, and a good cause to fight for. And still more, God is here among the hills of Gauley, as well as in the town of Marietta or at the home fireside."

It was decided that the fortifications previously constructed at Fayetteville were not adequate and during December and January they were strengthened considerably. It was at this time that the Federal Fort Toland was constructed. The walls were reportedly thirty feet wide at the base, with a large parapet which would allow a team and wagon to pass along its surface. Gun embrazures were placed in all directions, and the enclosure was intended to accommodate two or three regiments of troops, as well as provisions sufficient to garrison them for several months. Around the fort, abatis and ditches were placed, offering serious obstacles to any attempt at assault. The Yankees calculated that a force of three thousand men could successfully defend Fayetteville against four times that number of assailants.[4]

Among the civilians of Fayette County who were "sick and tired" of the war, were George and Nancy Hunt, who operated a store close to the James River and Kanawha Turnpike at Mountain Cove. The Hunts had moved to Fayette County from New York in the 1850s. They were members of a spiritualist colony established at Osborne Creek between 1850 and 1852. When the war began four of the Hunt boys dodged Confederate conscription and escaped into Ohio to enlist in the Federal army. Mrs. Hunt wrote to her friends in New York on December 7 from her home at Mountain Cove:

> I was so vexed with the Yankees I hoped they would never come back but I soon got tired of the Rebels and wished the Yankees would rain down by thousands. I got my wish. They have now been back 5 weeks but I have not yet seen one. The Rebels "done us mighty bad" while they were here. A company of guerillas headed by Bill Taylor came into our store and took possession in the name of the southern confederacy and intended to rob the store of everything, and did take about $100 worth of goods, but were prevented from taking more by the arrival of A. Forsythe who together with Sam Tyree, Bob Frazier and Bob Nichol who stopped them. They had no orders from the army to act thus and it was just a Bill Taylor raid . . . Jenkins Cavalry damaged us about $50 in one night when 300 or 400 of them camped on our place, burning rails and stealing as usual. Crow is one of them and is a fair specimen of the lot. I suppose they all feel as Parson Brownlow said he did when in Washington. The Parson said "I feel like I want to steal something." I felt all the time that the Rebels were here that they would. I tell you I did not rest easy while they were here. I was so afraid they would take my husband off again but I found we had a good many friends even in the Rebel army. . . . While the Confederates were here we sold our entire stock of goods, old rubbish and all and could have sold $100,000 worth if we had had them. I tell you there is nothing to be had in Dixie. I am glad we did

not have any more and wish we did not have as much as we did for such pay as we had to take. Confederate rags. There is no store nor any goods to be had this side Gauley Bridge and few can get in to go there. There are troops at the Tompkins place and they are very strict. Only now and then can one get through the picket line.[5]

During December and January two new camps were established in Fayette County. The largest of these was Camp Reynolds at Kanawha Falls, the other was Camp Vinton, which consisted of a few cabins and a livery stable near the mouth of Loup Creek, present-day Deepwater. Camp Reynolds was occupied by the 23rd Regiment Ohio Volunteer Infantry, commanded by Colonel Rutherford B. Hayes. At Camp Reynolds with Hayes was a young lieutenant, William McKinley. Thus the 23rd Ohio was the only Civil War regiment to boast two future presidents. Hayes entered the White House in 1877, McKinley in 1897.

With winter already at hand Colonel Hayes put his men to work quickly. They had to dig drainage ditches, lay-out a parade ground, and construct log huts. Hayes's cabin was built as a double, with two rooms eighteen by twenty feet each. When he moved in he found it somewhat less than comfortable. It had a "shake roof," shingled with large openings between the shingles, and the snow came through in clouds. Hayes said he felt he should sit before his fire with an umbrella over him.[6]

View of Kanawha Falls showing tents and wagon shop of the Union Army. This view was taken from the south side of the Kanawha River.
COURTESY AUBREY MUSICK, GAULEY BRIDGE, WV

View of Camp Reynolds, Kanawha Falls, January 1863.
COURTESY RUTHERFORD B. HAYES LIBRARY, FREMONT, OHIO

War period drawing of Colonel Hayes cabin at Camp Reynolds Kanawha Falls. COURTESY RUTHERFORD B. HAYES LIBRARY, FREMONT, OHIO

Picture of a mill on Kanawha Falls drawn by a soldier during the Civil War. COURTESY RUTHERFORD B. HAYES LIBRARY, FREMONT, OHIO

Federal soldiers on Van Bibbers Rock, Camp Reynolds, Kanawha Falls, 1862. COURTESY FAYETTE COUNTY HISTORICAL SOCIETY

The 92nd Ohio Infantry built Camp Vinton. Lieutenant George B. Turner wrote to his mother from there on Christmas eve 1862:

> I might give you our bill of fare for dinner tomorrow. Steak fried with thickened gravy, soft bread, butter, molasses, beans, mustard and a couple of pies perhaps, from the bakers. Not very bad, I think.... I have been superintending for several days the building of our (the Capts.) quarters. The cabin is about sixteen by eighteen within, with a kitchen attached. Yesterday the men finished putting on the shingles, which are oak and fastened with weight poles, as we can get no nails. We are to have a large chimney with an old fashioned fireplace, where we may lay in the old and almost forgotten luxury of a "back-log." I hope the buildings may be finished soon, as I am anxious to get the company drilling again.... The baker is turning out pies and cakes for Christmas today. They are bought readily by the boys and are a source of considerable profit to Henrick & Co.... I shall wish you all a "Happy New Year" as this will be too late for Christmas.[7]

The "Halfway House" or Tyree Inn at Ansted. This house was built around 1780 and was used during the winter of 1862-63 as headquarters of the Chicago Grey Dragoons. Today the site is privately owned.
AUTHOR'S COLLECTION

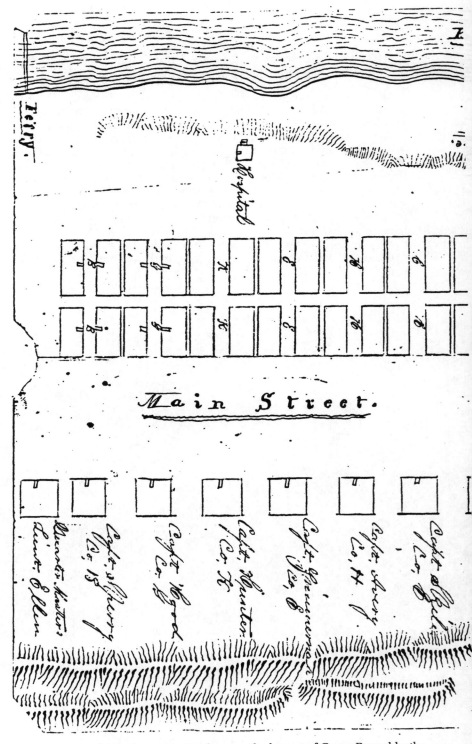

Map drawn January 1863 showing the lay-out of Camp Reynolds, the Federal camp at Kanawha Falls. COURTESY DON MINDEMANN

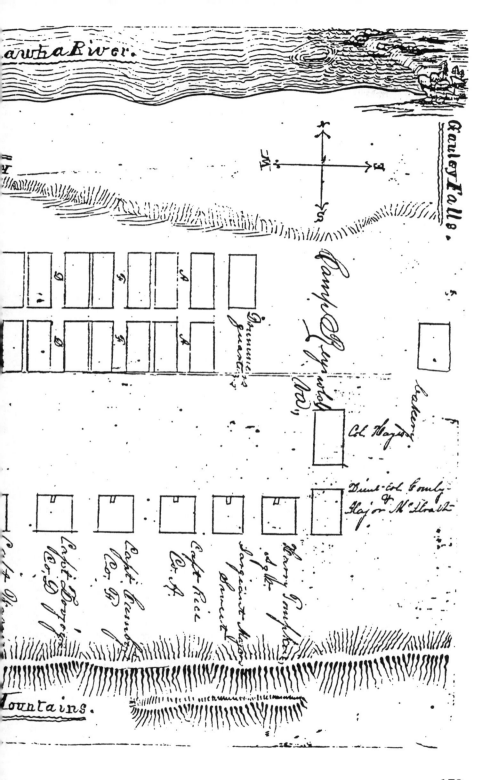

The following article, by correspondent "D" of the Second Virginia Cavalry, entitled "Fayetteville in War Time, 1862," appeared in the Ironton Ohio *Register* on December 18, 1862:

> When I last wrote we were domiciled at Gauley, probably the most God forsaken country on this green earth. Now we are quartered in and near the village of Fayetteville, the shire town of Fayette County. It is a beautiful location; part of the village is on high ground with here and there small houses in the valley. From all appearances, before the war, the villagers really lived at home and had their respective places of residence surrounded with beautiful trees and shrubs of every kind.
>
> The village before the arrival of our boys was almost entirely de-populated having only one family living in it. Most of the buildings were much mutilated by the soldiers leaving only two or three houses untouched, one of which is a new and spacious brick where General Scammon is now quartered with his staff and other high officials. Another house when we arrived here was used as a rebel hospital; now it is used by our forces as a hospital. All the formulae of war are observed by this portion of the Kanawha division, guard as well as picket duty.

Christmas day was observed at Camp Reynolds with a target shooting contest authorized by Colonel Hayes. Hayes offered a turkey for first prize, a bottle of wine and glass for second place, and a bottle of wine for third place. The day was thus spent in relaxing and joyful manner, as much as could be expected in war time. Hayes's troops had nearly completed their cabins by Christmas day and by early January everyone was comfortably housed. Each company had three cabins for the men and one for the officers, making a total of fifty-six cabins in all, including a separate drummer's quarters, bakery, and hospital. Hayes's cabin was at the eastern end of the camp and faced the road and parade ground which ran between the rows of cabins. Montgomerys Ferry was near the western extremity of the camp within easy reach for the loading and unloading of supplies.

Colonel Hayes preferred the active campaigns of the spring and summer, but now had resigned himself to a period of inactivity. Campaigning in the winter in the mountains was almost impossible, and anyway the Union forces in West Virginia were not large enough to undertake an advance that would get anywhere. Their available strength was diminished still more in January when General Crook's division was sent to reenforce the Federal army in Tennessee. With Crook's departure command of all the troops around Gauley Bridge and Fayetteville devolved upon General Scammon. On January 7 Colonel Hayes was placed in command of the First Brigade of the Kanawha Division, consisting of his own 23rd Ohio along with the 89th Ohio Infantry and two cavalry units, Captain Gilmores and Captain Harrison's, for a total of 1,859 troops. Due to sickness and desertions Hayes estimated that he actually commanded 1,350 troops.[8]

On January 24 Colonel Hayes's wife and two boys arrived at Camp Reynolds. They would stay with him for nearly two months. The boys rowed, fished, rode horseback, and enjoyed all the varied activities of an army camp. Mrs. Hayes, gracious and tactful and obviously aware of her obligations as an officer's wife, made a special effort to impress the regiment. She visited the hospital and the quarters and won the affection of the officers and enlisted men alike. A favorite camp tale was of the recruit who asked a comrade: "Where is the woman who sews on buttons?" The friend mischievously directed him to the house of the Colonel's lady. The soldier there accosted Lucy and inquired if she was the female who fixed the buttons of the regiment. Mrs. Hayes took in the situation and accepted the blouse the man offered. "I will be glad to do the work," she said, "leave it with me and I will see it is done." By the time he was to recover the garment he had discovered the truth and was terribly embarrassed upon his return. Mrs. Hayes assured him that it was all right, that she and he had turned the trick on the perpetrator.

During the winter months of 1863 General Cox and other high command officials were coming to the conclusion that West Virginia was not a very promising base from which to launch offensive operations. They were content to hold their present territory, which embraced the most valuable areas and which in June of 1863 became the new state of West Virginia. The Confederates, too, had just about written off any hope of recovering the region. They sought to maintain their positions along the New River Narrows to intercept any Federal advance on the Virginia and Tennessee Railroad.[9]

Thus the early months of 1863 were spent in camp, everyone looking forward to the warm weather and active campaigns of springtime. In the middle of March orders came for Hayes to garrison Charleston. He moved there with his regiment and went into quarters at Camp White across the river from the town. The other regiments of his brigade were camped at various posts from Gauley Bridge to the Kentucky line. Shortly after moving to Camp White, Hayes's force was augmented with the 5th and 13th Virginia Infantry, really composed of West Virginia Loyalists.

Hayes was not reassured by his reenforcements: "I am in command of some of the best and some of the poorest troops in service," he observed. He did not think that he could hold his position against a formidable attack.[10]

POSTS RETURNS FOR GAULEY BRIDGE AND TOMPKINS FARM
MARCH 1863

Commanding at Gauley Bridge—Captain Seth J. Simmonds
Commanding Artillery—Lieutenant D. W. Glasie
　　　　　　　　　　Simmonds Volunteer Battery
Commanding Infantry—Lieutenant R. Blazer
　　　　　　　　　　Detached Infantry, Companies B & G 91st
　　　　　　　　　　Regiment Ohio Volunteer Infantry

Total enlisted men present—150
Officers present—3
Horses serviceable—104
Horses unserviceable—9
Weapons on hand—Sharps Carbines .46 calibre—Springfield Muskets .58 calibre
Artillery—six 10-pound parrott guns
Commanding at Tompkins Farm—Sergeant Wm. E. Brand
　　　　　　　　　　Detachment—Company A
　　　　　　　　　　First Virginia Cavalry

Number of Companies—1
Total enlisted men present—15
Officers present—0
Horses serviceable—14
Horses unserviceable—4
Weapons on hand—Sharps Carbines .46 calibre
　　　　　　　　　　Springfield Muskets .58 calibre

Artillery—0

Total present at both camps—168

　　　　　　　　Certified by Captain Seth J. Simmonds—March 26, 1863[11]

■ *Chapter Ten*

From Fayetteville To Appomattox

As the season for resuming military operations in 1863 approached, it was apparent that the Federals were massing their strength for another advance toward Richmond. General Lee determined to delay and embarrass such an operation by striking at the Baltimore and Ohio Railroad, over which a large part of the supplies and reenforcements were sent to the Army of the Potomac. General William E. Jones and Brigadier General John D. Imboden were entrusted with this work. This became the famous Jones-Imboden raid, which set the stage for the last Confederate artillery attack on Fayetteville.

General Imboden left Shenandoah Mountain near Staunton, Virginia, with 3,400 men on April 20 1863. General Jones left Lacey Springs, Virginia, on April 21 with 1,300 men. Their raid carried them into a large section of northern West Virginia and then south by two separate routes. They were expected to unite at Summersville, Nicholas County. Fearing that the Yankees would send a regiment to Summersville to intercept them, General McCausland was directed to make a demonstration against Fayetteville, and prevent the Federals from detaching any of their force.

At that time, Colonel C. B. White, with the 12th Ohio, two companies of the 2nd West Virginia Cavalry, together with two sections of McMullin's Artillery Battery of six 3-inch rifled guns, was stationed at Fayetteville. McCausland had the 36th Virginia Infantry, six companies of the 60th Virginia Infantry, four pieces of Bryan's Artillery Battery, and a company of cavalry.[1]

On May 16 McCausland's troops moved out of camp toward Fayetteville. They skirmished with Federal pickets and cavalry along the road between Raleigh Court House and Fayetteville and appeared before the town at 2:00 p.m. May 19. Two artillery pieces from Bryan's Battery were quickly placed on a cleared ridge facing the Federal forts, and a third piece, commanded by Sergeant Milton W. Humphreys, was posted on a plateau at the end of a straight opening, which had been cut in the woods and ran directly toward the Federal fort. Humphrey's gun opened first and was immediately answered. The Yankees returned fire so vigorously and with such accuracy that Humphreys was compelled to move his cannon to a place nearby where they could be hidden by the timber in front and the smoke behind them from the woods, which were on fire. Once in place, Humphreys began shelling the Federals by firing over the intervening forest. This demonstration set a precedent for modern warfare by the use of indirect fire. An artillery duel was thus begun, which continued at intervals until evening.[2]

At daylight of May 20, two pieces of Bryan's Battery, which were situated about three hundred yards to the left of Humphrey's cannon, began the second day's duel. After about ten rounds had been spent, the Yankees replied. As the firing continued, W. S. Newton, a Federal surgeon stationed at Fayetteville, attempted to write a letter to his wife. He wrote:

> We are here, and the enemy in front on the Rolla road, they drove us in yesterday, and opened on us with three or four pieces of artillery. We replied with part of our guns, our men occupy the breast works, and have hardly returned their fire, saving our ammunition till it is needed. . . . We have had but few casualties on our part today, some yesterday however. . . . the shells burst very near us, yet we feel we have as good position as yet can be obtained. Ned is here with me and enjoys the excitement very much. They opened on us this morning at 4 a.m. and have kept it up almost constantly. We are hopeful and confident of holding the place. Give yourself no uneasiness for I shall take care of Ned. . . . I cannot write anything while the booming of cannon at every moment strokes the ear. . . . My regards to all the friends, much love to Matt & Kittie, in haste, your husband.[3]

The Confederates remained in front of Fayetteville until about 2:00 p.m. of May 20, at which time they withdrew. After several hours had passed Colonel White, of the 12th Ohio, received permission to follow. He started about dark with approximately two thousand men and a detachment of artillery. This pursuit resulted in an artillery fight at Raleigh Court House, after which McCausland's troops encamped a few miles south of Raleigh on Piney Creek. One of the Southern soldiers who had been with McCausland at Fayetteville was Gordon Thompson of the 60th Virginia Infantry. Thompson described their attack in a letter to his wife on May 24:

> Dear Wife I again take the opportunity of droping you a few lines to let you know that I am well at this time hoping when those few lines come to hand that they may find you all well. We have had a pretty severe march. We marched to Fayetteville and there we attacked the yankees and run them back into there fortifications and there we shelled them for 14 hours. They returned the fire briskly, we lay one night and part of two days under the range of their shells. We had 3 men wounded in all and none killed. The yankees loss was 2 or 3 killed and several wounded. We left Fayetteville and retreated back to Blakes and there stayed all

Camp of the 12th Reg't O.V.I. at Fayetteville Va. April 1863.
A. Fort Scammon. — B. Battery McMullan.— C. Camp of the 12 Reg't O.VI.— D. Fayetteville Courthouse.

COURTESY WVU ARCHIVES

night. The yanks followed us and run our pickets, us firing at them. We formed a line of battle and the yanks came in sight and we gave them a volly and they fell back out of our sight. Then we came on back to Raleigh Court House. We stopped at Beckleys in the field on this side of the Courthouse and stopped to stay all night. The yanks came to the top of the hill on the other side of the Courthouse and fired there cannon several times over towards the Courthouse trying to find out where we were. We left Beckley then and come to Pina. How long we will stay here I cant tell, we have taken some 10 or 15 prisoners and 10 or 15 horses. I have had my health very well since I left....I think the yankees has fell back from Raleigh at this time . . .[4]

There had been no infantry fighting in front of Fayetteville and hence scarcely any casualties. The second shot fired by the Rebels killed Sergeant Owen McGinnis of Company A, 12th Ohio Infantry. He was struck on the head with a ball from a twelve pound spherical case-shot, killing him almost immediately. The Yankees lost another man killed, seven wounded, and nine missing. McCausland's forces had several wounded and none killed.[5]

Sergeant Gordon Thompson, Company H, 60th Virginia Infantry. He fought in the battle of Fayetteville, WV, May 19, 1863.
COURTESY ROGER THOMPSON, HUNTINGTON, WV

The May 1863 attack on Fayetteville was the last one of its kind in Fayette County. Just one month later the new state of West Virginia was officially admitted to the Union, thereby severing forever trans-Allegheny Virginia from the Old Dominion. This unprecedented and somewhat questionable act prompted General Henry A. Wise to declare that Western Virginia had been torn from its moorings of state-hood and nation without rhyme or reason and was the "bastard offspring of political rape."[6]

West Virginia remained under Federal control throughout the balance of the war, and although Confederate attacks in this region continued they were more as a harassment than as a serious Southern attempt to control the region. The period between West Virginia statehood, which became official on June 20, 1863, and the end of the war, was rife with skirmishing attacks and raids into this region. Probably the first such attack into Fayette County occurred exactly one week later, June 27. A group of Confederate Partisan Rangers made a raid along Loup Creek to its mouth. At the Federal Camp Vinton they captured the dozen or so soldiers posted there and planned to burn the Yankee wharfboat and supplies. They had crept up on the Yankees in the middle of the night and when dawn broke had their captives on horses ready to move out. Directly across the Kanawha River from the Loup Creek landing was a large Union camp. Looking across the river in the early morning light the pickets could tell something was wrong, as described by W. S. Newton of the 91st Ohio Volunteer Infantry:

Daylight coming on, our Lieut. with his 20 men, could disern what was going on. He saw them all disarmed and mounted, ready to move with their prisoners. He ordered his men to fire on them across the river, which they done with such aim as to empty several saddles, and repeated it so often, as to throw them into confusion, at the same time called to the boys to dismount and take cover under the bank, which most of them did, and escaped. Crawford kept up such incessant fire, the Rebs were obliged to run, and leave the most of their prisoners, the horses too.... A party of Rebs, on a scout, also this morning, halted one of our trains at Mrs. Tompkins farm, and took from it fifteen mules, with which they made tracks. They are unusually active, all around us, and are showing any amount of daring.[7]

On July 1 General McCausland was again asked to make an advance on Fayetteville and report his estimate of the force that would be required to take the city. When his scouts reached Fayetteville, they discovered that the Union fortifications had been reenforced. McCausland reported that he would expect an attack on his forces rather than the other way around. After remaining in the vicinity briefly the Rebels withdrew, having had only a light skirmish with the enemy.[8]

On July 7 a raid was conducted by Rebel cavalry along the Cotton Hill road. Captain J. V. Young, of Company G, 13th Virginia Infantry (US), described the event in a letter to his wife: "On the 7th of this month the Rebs attacked our trains on Cotton Hill and took all the teamsters prisoners, and the best of the horses, and made their escape. There are several companies, or bands, of horse thieves and bushwackers in this county, and they are very trouble-some to our transportation. I am willing to deed all this country to Jeff Davis if he will stop his 'Gray Backs' from stealing, and if he doesnt stop them I reckon we will have to do it, or stop them from breathing, but we will have to catch them first."[9]

During early August the 13th Virginia Infantry (US) were moved from their camp at the eastern foot of Cotton Hill to a new position at Sewell Mountain. They occupied for several weeks "Camp Defiance," the 1861 camp of General Wise. This move was apparently favorable to the men, as described by Captain Young: "The whole regiment is here, and camped where the Rebels camped two years ago. It is such a nice place, cool and shady. I hope we will remain here some time."

With no active campaigning taking place in this region, the Federals spent much of their time fortifying their posts against attack. The works at Fayetteville had now become quite extensive and made a lasting impression on W. S. Newton, of the 91st O. V. I.:

> If this war ever ends, this place will be visited for years to come, as a place noted for its defences. You, or anyone, not acquainted with the fact, would be surprised at the extent of the works here. They will stand as a lasting monument of this wicked rebellion. Whether they will be noted for any serious engagements, or long sieges, is a matter of some

Major Rutherford B. Hayes. This future 19th President of the United States served extensively in West Virginia as a member of the 23rd Ohio Volunteer Infantry. COURTESY RUTHERFORD B. HAYES LIBRARY, FREMONT, OHIO

Lieutenant William McKinley. As a member of the 23rd OVI he spent much of his military career in West Virginia. He became the 25th President of the United States.
COURTESY RUTHERFORD B. HAYES LIBRARY, FREMONT, OHIO

doubt in my mind. We can, it is estimated, resist an attack of ten or twelve thousand, and do it successfully. This place is where Col Jack Toland showed his fighting qualities, about a year ago. He had two horses shot under him, and then on foot, led his men in thickest of the fight.[10]

On September 11, there was a brief skirmish at Gauley Bridge between Union pickets and Southern scouts, and in early November there was another raid on Camp Vinton, as described by Captain Joel H. Abbott:

About the first of November our cavalry was left in Raleigh to guard the road and to use some forage on the Ferguson and other neighboring farms. Captain Irvin Lewis with part of his three companies made a raid down lower Loop Creek to its mouth, surprised and captured sixty-five men and seventy-five horses and equipment. We retreated out up Armstrong Creek, crossed over Payne's Mountain, where we rested and ate up everything we had. We went down a ridge to Paint Creek and then to Raleigh court house. The prisoners were sent to Dublin Depot.[11]

In February of 1864 General George Crook was detached from the Department of the Cumberland and assigned to the command of the Third Division, Department of West Virginia, then in the Kanawha Valley.[12] The Federal forces at Fayetteville began their spring campaign in March with a reconnaissance to Summersville. They found no enemy there and so returned to Fayetteville.

General U. S. Grant had planned a general movement to start about the first of May, and General Crook's forces were to move from the Kanawha region through Lewisburg; Colonel Thomas M. Harris was to move from Beverly, and General Franz Sigel was to move up the Shenandoah Valley and join Generals Crook and Averell near Staunton. The purpose of the whole movement was a raid on Dublin Depot and the New River Bridge on the Virginia and Tennessee Railroad. The destruction of the railroad would cut the communications of Generals James Longstreet and Sam Jones in east Tennessee, and possibly compel the Confederates to abandon that country. It would be impossible to reconstruct the railroad during the campaign and might result in the evacuation of Richmond.

The supplies and transportation for this great movement were accumulated at Fayetteville during February. On May 1, General Crook ordered the forces there to be ready to move at a moment's notice. On May 2, the other regiments and artillery began to arrive.[13] Captain J. H. Prather of the 91st Ohio Infantry described the scene as orders to move were received: "The long roll sounded through General Crook's camp at Fayetteville, on the morning of May 3rd, 1864, and was greeted by thunderous applause from the throats of 7,000 Union soldiers, and perhaps in little less than an hour the column was in motion headed for south West Virginia, via Raleigh, Flat Top Mountain, Princeton and Shannon's roads."[14]

As the long Federal column marched out of Fayetteville they left behind many of their comrades who were sick and in the hospital. A large detachment of infantry was left to hold the town and everyone's attention focused on the great raid. This raid was successful for the Federals, and after marching on Lynchburg, the army marched back down the Kanawha Valley. This was the last large scale movement of troops through this region, and it signaled the resumption of active campaigning for 1864.

The increase in military traffic along the James River and Kanawha Turnpike was described by Nancy Hunt of Mountain Cove. She wrote:

> It seems very much like when General Rosecrans was up on Sewell Saturday week the longest train of wagons passed here that I ever saw. A regiment was along guarding 262 prisoners and 450 contraband and a long train of ambulances containing the wounded. Trains and soldiers are passing daily, 25 prisoners, some of them bushwackers passed today. . . . Some 8 or 10 of them had been shot. A paper is pinned to them with inscription, "this is the fate of all bushwackers." Much bushwacking has been going on on Sewell Mountain. A soldier and his wife took supper here a few nights ago, who were fired on from the brush this side of Frank Tyree's and he was wounded in the leg. I suppose the forces at Meadow Bluff and Lewisburg will start another raid tomorrow toward Richmond as the trains passing here are hurrying up. I have been busy waiting on soldiers with nobody to help me....I am alone most of the time. Soldiers behave very well generally and pay well though some of them will steal. One stole a large silver spoon from me yesterday. I followed the soldiers down to Piggots place where they were camped but did not get my spoon. The officers said they would watch and try and get it for me. There is a telegraph wire stretched before my door which looks a little like civilization.[15]

POST RETURN FOR FAYETTEVILLE, WV.
JULY 1864

Commanding at Fayetteville — Colonel Harvey Crampton
146th Regiment
Ohio Volunteer Infantry

Number of Companies — 10
Total enlisted men present — 854
Officers present — 36
Detachment of 7th Virginia Cavalry — Commanded by
Lieutenant Webb

Number of Companies — 1
Total enlisted men present — 25
Officers present — 1
Horses serviceable — 25

Horses unserviceble—0
146th OVI Company Commanders—
Company A—Captain S. B. Morris
Company B—Captain J. D. Hendrickson
Company C—Captain R. H. Williamson
Company D—Captain A. Miller
Company E—Captain O. H. Denise
Company F—Lieutenant J. R. Whitacre
Company G—Captain W. G. Foster
Company H—Captain D. Weidner
Company J—Captain A. Brown
Company K—Captain S. C. Keever
Field and Staff Officers 146th OVI—
Lt. Col. John R. Hitesman
Adjt. Fred S. VanHarlingen
Qmst. Ionas W. Stubly
Surgeon—Isaac L. Drake
1st. Asst. Surgeon—Otho Evans
2nd. Asst. Surgeon—Amos Sellers
Chaplain—Adolphus S. Dudley [16]

On July 4 1864, occurred the last organized Confederate attack on Fayetteville. Captain Joel H. Abbott of the 8th Virginia Cavalry, advanced from Princeton, Mercer County, and raided the town, as he later described: "On July 4th, I raided Fayetteville and captured four sutler wagons and a large quantity of all kinds of goods, and carried the goods out on our horses. These wagons were placed outside of the lines for the purpose of trading with the people on that day."[17]

The remainder of 1864 and the tumultuous spring of 1865 passed without any major military activity in Fayette County. The actions of Southern Partisan Rangers and bushwhackers constituted the only resistance offered the Yankees in this county during the closing months of the war. With the surrender of General Robert E. Lee at Appomattox, Virginia, on April 9, 1865, Fayette County was once again "working alive" with Confederates. This time they came not as liberators or invaders, but simply as Confederate veterans. Many of them had fought since the earliest days of the war and now returned home to "pick up the pieces" of their lives. Still others would never return for they had been killed on the field of battle, or had succumbed to disease.

After four long years of war, it was difficult for many people to believe the end was now at hand. On April 23, 1865, Nancy Hunt expressed her cautious optimism in a letter to her friends: "Can it be true there will be no more fighting? Shall I see no more armies pass here? Nary another Reb? No more fear of guerillas? This is too much to take in all at once! I hope it is all true."

■ *Appendix*

The Dixie Rifles

My First Ninety Days, or the Blunders of a Confederate Captain

As told by Colonel B. H. Jones

About the middle of June 1861, I raised a company of infantry in Fayette county, then a part of Virginia, and was elected Captain, but certainly not on account of my familiarity with the pages of Scott, Gilham, or Hardee, as the sequel demonstrated, for I "had never set a squadron in the field, nor the division of battle knew more than a spinster."

In the latter part of the same month, the company, glorying in the euphonious and significant appellation of the "Dixie Rifles," was regularly mustered into the service of the Confederate States at the great Falls of Kanawha, by Brig. Gen. Henry A. Wise, to whose "legion" it was attached.

I had just returned from Lewisburg, and sported a gray jacket, gotten up by a tailor of that place, who, by way of securing the job, had assured me that he was perfectly "Au fait" in all the minutiae pertaining to the decoration of military rank.

I was quite proud of my up-buttoned, close-fitting "Jacket of Gray," and felt all the importance of the commander, until I was startled from my dream of consequentiality by being addressed by an old soldier as "Corporal Jones." My "Knight of the Shears," equally ignorant with myself had braided me a corporal. My mortification was excessive, nor did I recover my usual composure until spasmodically I tore off the libelous braid, and cast it disdainfully upon the ground.

In the afternoon of that day it became necessary to draw rations and as our supply was at Gauley Bridge, two miles above our encampment and no transportation at hand, I was under the necessity of marching the men up, so they might carry down their "hard tack and bacon." Ignorant of the command necessary to form two ranks, or even to face them in the direction I wished to move, I took my orderly sergeant aside, communicated my intention to move at once upon the supply depot, and directed him to form the company in single line, with the men facing toward the bridge. He thought the suggestion a happy one, and proceeded to execute the order, by taking each man by the jacket collar and forcibly establishing him in the proper position, always accompanied with the important injunction to "stand right there."

At the command, "forward march," given with all the energy I could summon to the aid of a pair of strong lungs, the "Dixie Rifles" moved off in the most approved style. The interminable lines winding with the frequent curves and angles of the road, which coupled with the irregular and unrestrained swinging to and fro, from right to left, and from left to right, of one hundred and eighty awkward arms, brought forcibly to the mind the spiral and confused locomotion of a mighty centipede. Ever and anon, reaching a commanding point, I would cast backward a glance of pride and satisfaction at the vast proportions of my command. Caesar, Alexander, and Napoleon, at some period in their eventful lives, possibly, felt as well as I did then, but I will never concede that either of them ever felt any better. There was but one unpleasant drop in the cup of my happiness. I knew that the company ought to march in two ranks, but how to get it into that shape was the rub.

I had warned the company on taking up the line of march, that "talking" in ranks would not only be highly unmilitary, but could not be tolerated at all; so that not a sound broke the funeral silence of that two-mile march, save once when an old soldier, who had seen service in Mexico, ventured to speak in a subdued tone to the man immediately in front of him. I detected this, and jealous of my authority as well as being indignant at so wanton a breach of military propriety and stung by what I suspected was a merited criticism upon the Indian file movement and consequently a reflection upon my military accomplishments, I sternly ordered him to be silent, reminding him that I was captain, and as such not to be trifled with, and that as an old soldier he should know better.

At length we arrived at Gauley Bridge; the rations were issued, and the order, "shoulder bacon and hard tack," was about to be given, when as luck would have it, up came a four horse wagon driven by a Nicholas county farmer. Fancying myself, by virtue of my captaincy, vested with extraordinary power, in other words a gentleman of "high claims and terrifying exactions," I proceeded at once to press into service wagon, team and teamster. The farmer protested, alleging that he had been long from home, and could not reach there until late at night, but all this was unavailing. He had encountered what was afterward known as "military necessity," and as a matter of course had to succumb.

Now another difficulty stared me in the face. My men had fallen in line, facing Gauley Bridge and I wished to move them in the opposite direction but I did not know the command for a counter march. "File right" and "File left" were terms unknown to me, or if known, were utterly meaningless. I reflected a moment, nervously twirling my cane, for sword I had none, my face burning, my heart beating audibly, the men silent and expectant, until finally growing desperate, I cried out in the extremity of my agony, "Men, turn your faces the other way." Some turned to the right and some turned to the left, some made the entire circle and stood as at first, while others with countenances as blank as lamp posts, made no effort whatever to obey the order, while my sharpened hearing caught a half-suppressed sound of malicious laughter in the direction of the "old soldier." Finally with the aid of my orderly I got them all turned right and the interminable line that "Like a wounded snake, dragged its slow length along," returned to camp. Here the wagon of the Nicholas county farmer was speedily unloaded. The farmer then approached me in the most deferential, not to say awe-stricken manner, and stammered something about pay. I was astonished at his ignorance of the license of military authority, and indignant at his want of patriotism, replied with much spirit, "Pay, sir, pay indeed. No pay at all, sir. A mere gratuity that you as a loyal and patriotic citizen should esteem a privilege to render to your country, sir: the Southern Confederacy, sir." With alarm, wonder, mortification and disappointment all depicted in his countenance, he shrank back, took up the lines, cracked his whip, and was soon out of sight. He should have thrashed me soundly on the spot, although at the time, I honestly believed that I was merely exercising an official prerogative for the benefit of my country.

A happy idea now suggested itself. I would solve the vexed problem of forming a company into two ranks by being present when Captain Riggs' company was on parade. His men were formed in line. "Facing by the left flank, in two ranks, form company. Company, right face. March" said Captain Riggs. "Eureka!" I almost audibly ejaculated; then hurriedly dodging around the corner of an old house standing close by, I hastily took my memorandum book from a side pocket, and eagerly recorded with pencil the talismanic words. By roll call next morning I had memorized them, and was enabled to accomplish the wonderful evolution of forming a company in two ranks to my own entire satisfaction. As to four ranks, I had never heard of such a thing, and should have been strongly inclined to question the sanity of the man who would have mustered with the "melish" nor seen the inside of any work on military tactics. Mine was not an isolated case; it was the experience of nine-tenths of the Confederate officers. We were green, yes all of us—succulently green.

The enlisted men drew rations regularly, but when I applied for mine, I was politely informed, much to my mortification, that "rations" were not issued to commissioned officers. "How am I to live, sir? I anxiously inquired. "Indeed, captain, I am not able to answer your questions, though I would be most happy to do so, for it is one in which you are much interested," replied the commissary, who was as ignorant of his duties as I was of mine, though not in half so much danger of starvation.

I had not been long at Gauley Bridge until it came my turn to act as "officer of the day" and I felt both complimented and alarmed. I was wholly ignorant of the duties of my position. Major Bradfute Warwick, who subsequently, as a colonel, fell covered with glory at Cold Harbor, was commandant of the post. He was an eastern man, and unacquainted with the geography of that part of Virginia and of the disposition of the inhabitants, and so fancied we were in constant danger of a surprise. In this he was most energetically seconded and sustained by Captain Buckholtz, an officer of much gallantry who was in command of the artillery. He instructed me to visit the pickets twice during the day and three times during the night. The distance to be traveled in making the rounds was about six miles. The roughness of the route, and the labor and peril to be encountered, can not be conceived by any one that has never experienced such a task. In addition to the regular pickets, men were stationed about seventy yards apart, connecting each post with the main camp. These men being perfectly "green" fancied a live Yankee under every bush, and were ready to fire at the least noise.

I was to start on my rounds at precisely nine o'clock. I gathered by blanket and repaired to the guard house. It had rained almost incessantly for "forty days and nights," consequently the rude floor of the guard house bore a strong resemblance to that of a pig sty. I scraped away the looser particles of mud, however, and spreading my blanket, lay down among guards, Union prisoners, & e., but I consoled myself in my novel and uncomfortable position by the reflection that I was "serving my country," and that our forefathers had trod the same rugged pathway to glory and independence. Indeed my patriotism so far triumphed over my discomforts as to enable me to discover new beauties in the sentiment, "Dule et decorum est pro patria mori."

At the designated hour I was aroused by the officer of the guard, and then began the toil of the night. Varied and startling were the receptions and experiences of my dreary rounds. Sometimes it was a sharp "Halt! Who comes there?" Again it was a hesitation and nervous "Who's that?" And not unfrequently it was the startling "click" of a lock that made each particular hair stand on end as the aroused sentinel made ready to fire. I was under the necessity of specially instructing each sentry. When I got back to the guard house I found it was just midnight, the hour for "Grand Rounds," so accompanied by a sergeant and three privates I started again. My experience was about the same as on the preceding round, with one or two ludicrous variations. One sentry called out, "Who comes there?" 'Grand Rounds." was the reply, and this brought the response, "Come on, Grand Rounds." Another asked, "Who are you?" "Grand Rounds" we replied. "Oh, pshaw!" the sentry returned, "I thought it was them fellers coming to relieve me." With Grand Rounds completed, I found myself at the starting point at 3 A.M. wet, muddy and fatigued. I had only made two rounds and my instructions were to make three. To fail was in my opinion, death, ignominious death. I thought of my family which would be left in such unsettled and troublesome times without a provider or protector, and remembering how ill fitted I was for death, I nerved myself for the third round. Away I trudged all alone, and finished the third round about sunrise; the third round finished me about the same hour. I had traveled about eighteen miles, stubbing my feet against stones, falling over logs, jamming against stumps, splashing into mud holes, and wading a sluice of water no less than three times, thirty yards wide and three feet deep. As there were but three captains at a post, this task devolved upon me every third day. No wonder that my countenance grew haggard and wan, and my body weak and trembly, so that my own wife recognized me with difficulty. An iron man could not have endured such hardships unaffected. Yet I bore all cheerfully, and with martyr-like resignation and from a sense of duty, and because I thought at that time that the authority of a commanding officer was unrestricted, and that as a subordinate I was bound to obey all orders whether reasonable or unreasonable.

When commissioned officers were so verdant, what could have been expected of the private soldiers? Passing from my quarters to the creek one morning after sunrise to indulge in my usual ablution, I was suddenly halted by a sentinel, some fifty yards to my right.

"Who comes there?" he fiercely demanded.

"Captain Jones."

"Give the countersign, Captain Jones."

"The countersign is not required at this hour."

"Yes it is. Give me the countersign," he screamed, cocking his long mountain rifle, and bringing it to bear directly between my eyes so that I fancied I could almost see the bullet that was to finish my mortal career. It would have been imprudent to have shouted the countersign, surrounded as I was by laurel where an enemy might have been concealed, so I assayed to draw a little closer.

"Halt! Give the countersign, I tell you, or I will fire."

"I must get closer for I might be overheard."

"Don't care a darn. That's what the fellow said that came around last night, and I'm not going to fool with you much longer neither, Captain. So just sing her out."

Seeing that further parley or expostulation was not only useless but positively perilous, I yelled out "Jeff Davis."

"That's the truck, captain. Hurrah for Jeff Davis! You can go now," and he resumed his beat.

So apprehensive of a surprise were the superior officers—all eastern men—that finally I got badly scared myself. I was called up one night by Sergeant Major Pierce who informed me that the enemy were actually crossing New River just above the mouth of Gauley, in great force-using three flat boats for that purpose. I was ordered to awaken my men and get them under arms. The boys sprang eagerly to their guns, all but one poor fellow who was unfortunately at that moment attacked with violent pains in the region of the stomach. I was assigned a position in an oat field. Captain Buckholtz labored with great energy to get his artillery in position; wheels creaking, whips cracking, drivers swearing—there we stood in the field, shivering in the dark morning fog, with guns cocked, heads inclined eagerly forward, and eyes strained, vainly endeavoring to peer into the darkness; but no enemy came.

In the latter part of July we began the famous retreat—or as General Wise persisted in calling it, the "retrograde" movement—from the Kanawha Valley. Cox had been whipped at Scary; but another army acting in concert with his movement, was seeking by way of Sutton, Summersville, and the Wilderness road, to gain our rear and thus not only cut us off from our base but capture our whole force. Such confusion and demoralization as then ensued have been seldom witnessed. One entire company, perhaps two, deliberately filed off and went home. Another scattered like frightened sheep; but the captain marched boldly on along until becoming thoroughly disgusted, he broke his sword and wore his bars no more. Huge sides of bacon were pitched into the mud and trampled under foot. The heads of whiskey and molasses barrels were knocked in, and every man helped himself. The Gauley Bridge that had cost $30,000 was burned although the river was fordable for infantry and cavalry about one hundred yards above. It was said, though I never credited the report, that the famous "Hawks Nest" was examined with an eye for its destruction, but was declared non-combustible, and was thus saved for the admiration of future tourists. Every man went it on his own hook. For the first twelve hours, despite the efforts of the general, orders were disregarded and system was lacking. Quartermasters were oblivious to obligations to furnish transportation. My company baggage had been carried across the ill-fated bridge by the men. I waited for transportation until near nightfall, and the bridge was already in flames, lighting the heavens from horizon to zenith with the lurid glare. The army had gone, its retreating footsteps echoing amid the gorges of Gauley mountain, still no transportation, nor would the last lingering quartermaster answer satisfactorily my inquiries. So I moved off, leaving all behind.

Darkness soon came on, and the rain descended only as it did in the summer and autumn of 1861 in the mountains of West Virginia. We had gone a mile when we met a wagon with a four mule team and a negro driver. I pressed into service the wagon, mules and driver, and sent them back and got my baggage. It was now so dark that we could only with difficulty keep the road. I halted the company, told the men to take care of themselves, and they scattered in every direction seeking shelter under rocks and trees from the pitiless storm. I crept down the hillside, carefully feeling my way, and found a dry spot under a huge rock. I called to one of my lieutenants, who soon joined me, and I told him to go in first; he did as I directed, and so completely occupied the whole of the space that I was compelled to lie all night in the rain. While sleep to me was impossible, I could hear the lieutenant snoring boisterously all the night. I have never before regretted being generous, but I did that night, and I think you will concede I had reason. When morning came, I was wet to the bone, and chilled to the marrow. We started again at daylight. General Wise was standing at the top of Gauley mountain. When we came up, he told me to halt my company. I did so, and he furloughed every man who wished to go by his home. The result was, I entered Lewisburg with ten men out of ninety. I was so emaciated and careworn that my most intimate friends recognized me with difficulty.

Having recruited and reorganized his command, General Wise again advanced toward Gauley, while General Floyd, taking the Sunday road, moved towards Summersville, Wise encountered Cox at Big Creek, a few miles beyond the Hawks Nest, and, after a brisk skirmish, kept up until evening, owing to the failure of a part of his programme, fell back four miles to Woodville (the early name for Ansted), and went into camp. Floyd, having crossed Gauley at Carnifax Ferry, and entrenched himself on the cliffs, awaited Rosecrans, who, advancing from Cheat Mountain, attacked him with great impetuosity. The Confederates, though outnumbering at least six to one, resisted successfully every attempt to carry their position, until night-fall, when they withdrew with so much secrecy, that the Federal commander received the information of their retreat about sunrise the next morning, when his troops stormed and carried the undefended entrenchments. Floyd did not lose a single man in this battle but two were slightly wounded, one of them the general himself. Rosecrans must have suffered severely, as his men repeatedly assailed the works with great and persevering gallantry. Having recrossed Gauley, Floyd fell back to Dogwood Gap, at the junction of Sunday road with the James River and Kanawha Turnpike. Here he was rejoined by Wise. Cox and Rosecrans continued to advance with a force of at least 15,000 men, while that of the Confederates did not exceed 3,500. The latter fell back slowly to the top of Big Sewell Mountain.

Halting here a day or so, Floyd began to fortify, but suddenly changing his mind, he retired toward Meadow River, in Greenbrier County, and ordered Wise to follow, which he positively and indignantly refused to do, avowing his determination to oppose his 1,500 men against the 15,000 men of Rosecrans, and thus make a Thermopylae of Sewell Mountain. He accordingly named his position "Camp Defiance," and cooly awaited the advance of the enemy.

It is not necessary to recount how, in a few days, Rosecrans came up and went into camp in an equally favorable position about a mile west of "Camp Defiance;" how General Lee came down from the Cheat mountain region, bringing reenforcements; how he examined Wise's position, approved his course, and ordered Floyd to return; how for three or four weeks the rival hosts surveyed each other from their respective mountain strongholds without coming to an engagement; how, confident of the issue, we, from day to day, prayed for the advance of the Federals; and how, finally, one bright Sunday morning, when General Lee, as we were assured, had made up his mind to execute on the following Tuesday a great strategic movement that promised to result in the complete discomfiture of the foe, we got up and found he had struck his tents and precipitately retreated. I merely wish, in this connection, to relate one other of my adventures as a Confederate captain.

I was officer of the day. We were expecting Rosecrans to attack us. I wished to see the old officer of the guard. I did not know who had acted in that capacity the day before, and had not been enabled to find out by inquiry. I went to the guard tent and mentioned that I wished to see the old officer of the guard. A drummer suggested that if he were to sound his drum, perhaps the individual wanted would come to the guard tent. I cannot understand why I thought the beating of the drum would produce such a result, or why the drummer thought so. I know, however, I caught at the suggestion, and when the drummer asked me if he should sound the "long roll" I answered affirmatively, adding that I supposed that would do as well as anything else. The fact was I had not heard of the long roll before, and did not know that special significance attached to it. He commenced beating the long roll. There happened to be one or two officers in camp who knew that the long roll was a call to arms to repel an attack, or something of that nature. In other words, it was an alarm. They seized their swords, sprang out and called on their men to fall in, instantly; other officers caught the infection and followed the example. In a moment the entire camp was in an uproar, rivaling that of Babel itself. From every tent officers buckling on their swords, and privates with cartridge box in one hand and musket in the other, streamed forth like angry bees from so many hives, while above all other sounds, were heard the excited commands of officers, "Fall in, men, fall in!" "Back on the left!" "Out a little in the center!" "There, steady, front!" "Right dress!', &c., &c.,

Had I been capable, at that moment, of remembering anything at all that I had ever read, it would certainly have been Byron's "Waterloo":

"Ah! then and there was hurrying to and fro,
And there was mounting in hot haste; the steed,
The mustering squadron, and the clattering car,
Went pouring forward with impetuous speed,
And swiftly forming in the ranks of war;
And the deep thunder peal on peal afar;
And near the beat of the alarming drum,
Roused up the soldier e'er the morning star;
While thronged the citizens, with terror dumb,
Or whispering, with white lips: "The foe! they come,
they come'!"

But cries of "the long roll!" 'the long roll!" arising on all sides, assured me that I was the author of the mischief, and in the extremity of my mortification, I was senseless and dumb; and then Colonel Spaulding came rushing from his quarters, calling for his horse, and demanding, in an excited manner, "What does all this mean?" If ever man desired "A lodge in some vast wilderness—some boundless contiquity of shade," deep, dark, impenetrable shade, at that —one of the Bengal jungle variety—I was certainly that man.

In reply to his question, I succeeded, by a desperate effort, in stammering out that there was nothing serious the matter; that I had told the drummer to beat for the old officer of the guard, and he had with my sanction, beat the "long roll"—I being ignorant of the peculiar import and probable effect thereof.

For a moment, anger and a keen sense of the ridiculous appeared to struggle for the mastery; but the latter triumphed, and directing his orderly to tell the captains to dismiss their men, with an emphatic smile on his countenance, he invited me to his tent, and there good-humoredly explained to me the mysteries of the "long roll." Brave, accomplished, generous Spaulding! Two weeks later his body, a bloody corpse, was borne in a blanket to camp by four of his men. He had approached too close to the pickets of the enemy, and received two balls through his breast.

A few days after the retreat of the Federals from Sewell, General Lee sent two famous scouts—one of them afterwards Captain William Heffner, who was killed at the battle of Lewisburg, in May 1862. They were ordered to leave their guns in camp, as the object was information as to the location of the enemy. They found the enemy encamped in a field belonging to Colonel George Alderson. Under cover of the brush the scouts crept up to the fence enclosing the field, and while lying there, General Rosecrans and Cox rode up to within thirty yards, halted, and sat on their horses engaged in conversation for some time. Captain Heffner told me he could have counted the buttons on their coats. Had the scouts carried their guns, the career of the two Federal commanders would have ended that bright October morning—William Heffner and comrade were dead shots, with their long mountain rifles, at two hundred paces.

Suffice it to say, that I afterwards saw much service, and endured much suffering, for I was in the field from the beginning to the close of the war, excepting from the 5th of June 1864, to June 19, 1865, during which time I was a prisoner of war at Johnson's Island. I was with Lee, in the swamps of South Carolina, on the sand hills of Wilmington, in front of McDowell, at Fredericksburg; in the "Seven Days Battles," on the Chickahominy. With Jackson, at Cedar mountain; with Loring, in the Kanawha Valley; with Ranson, in the southwest; with "Tiger John" Mc-Causeland, at Piney, Princeton, and the Narrows; with Jenkins, at Cloyd's farm; and William E. Jones, at fatal Piedmont; but during those first "Ninety days" with Wise, in the Kanawha Valley and on Sewell mountain, I underwent more real suffering and hardship, than in all after military life.

And the "Dixie Rifles"; where are they now? Alas! some are sleeping beneath the magnolias of the south; some on the hills of Fredericksburg; some at Mechanicsville, Cold Harbor, and Frazier's farm; some at Piney. Princeton and the Narrows; some at Cloyd's farm; some at Peidmont, Winchester, Kernstown, Cedar Creek, Fisher's hill, and on the banks of the Opequon; some at the White Sulphur, Richmond, and Lynchburg; some at Camps Morton and Chase; some at Point Lookout and Elmira; some have gone home with broken constitutions; some maimed and almost helpless for life. With their gallant comrades of the glorious "old 60th," they everywhere bore their full share of suffering, and danger, and death; and, when at the close of the war, they, with streaming eyes and aching hearts, turned away from the "Conquered Banner," which,
"though gory,
yet is wreathed around with glory,
And will live in song and story,
   Though its folds are in the dust;
for its fame on brightest pages,
Penned by poets and all sages,
Shall go sounding down through ages
   Furl its fold though now we must."

In that sad hour, not more than a dozen of the original Dixie Rifles* answered at roll-call.

"On Fame's eternal camping round,
  Their silent tents are spread:
While Glory guards with solemn round
  The bivouac of the dead!"

Confederate memorial service conducted by members of the 36th Virginia Infantry reenactment group on June 27, 1987 at the grave of Stonewall Jackson's mother, Westlake Cemetery, Ansted, Fayette County.
AUTHOR'S COLLECTION

NOTES TO CHAPTER 1

1. J.T. Peters & H.B. Carden, *History of Fayette County, WV.*, (Jarrett Printing Company, Chas., WV. 1926) p.213
2. *Kanawha Valley Star*, WV. State Archives, issue of May 7, 1861.
3. United States War Department, *War of the Rebellion: A Compilation of the Official Records of the Union and Confederate Armies*, 70 vols. in 128 books, (Washington: Government Printing Office, 1881-1901), Series 1, V. 51 pt. 2: 788.
4. Ibid., p.791
5. Ellen W. Tompkins (ed.), "The Colonel's Lady: Some Letters of Ellen W.Tompkins: July-December 1861," *Virginia Magazine of History and Biography*, Oct., 1960 #69, p.387-419
6. Peters & Carden, p. 215
7. *Kanawha Valley Star*, issue of June 4, 1861
8. Boyd Stutler, *The Civil War in West Virginia*, Education Foundation, Inc., Charleston, WV., 1963, p.54
9. Letters Received 1861, V.M.I. Archives (Patton to Smith, June 4, 1861).
10. Official Records, V. 2 p. 51
11. Diary of Victoria Hansford Teays, transcript courtesy of Bill Wintz, Upper Vandalia Historical Society, P.O. Box 517, Poca, WV. 25159
12. Official Records, V. 2 p. 48
13. William Bahlmann, "Down in the Ranks," *Journal of the Greenbrier Historical Society*, Oct. 1970, V. 2 No. 2 p.43
14. Terry Lowry, *The Battle of Scary Creek*, (Pictorial Histories Publishing Co., Chas., WV., 1982) p. 45, Also Stutler, p. 50
15. Official Records, V. 2 p. 908-909
16. John S. Wise, *End of an Era*, (A.S. Barnes & Co., Inc., New York, 1965) p. 177
17. Tompkins Family Papers, Virginia Historical Society, Richmond, Va.
18. Wise, *End of an Era*, p.177
19. Beuhring H. Jones, "My First Thirty Days Experience as a Captain." *Southern Literary Messenger*, V. 37 No. 2, 1863.
20. Peters & Carden, p. 215
21. Confederate Records Group 109, Records of the Army of the Kanawha, Letters sent by General Henry A. Wise, June 1861-August 1864, National Archives.
22. Stutler, p. 52 – Note: Several historians have recorded that Col. Anderson accompanied Wise to Charleston; actually Anderson did not join the Wise Legion until July 12, 1861. See, National Archives, War Department Collection of Confederate Records, record group 109, Compiled Service Records of Soldiers who Served in Organizations from the State of Virginia. Records of Col. St. George Croghan, letter of July 12, 1861, Croghan to Wise.
23. Official Records, V. 2 p. 951.
24. Lowry, *Battle of Scary Creek*, p. 22
25. Official Records, V. 2 p. 197

26. Jacob Dolson Cox, *Military Reminiscenses*, (New York: Charles Scribner & Sons, 1900), V. 1, p. 59-60

27. Wise Letters, CRG 109, National Archives

28. War Department Collection of Confederate Records, Record Group 109, CSR, State of Virginia, records of Col. St. George Croghan.

NOTES TO CHAPTER 2

1. Boyd Stutler, *The Civil War in West Virginia*, Education Foundation, Inc., Charleston, WV., 1963, p.67

2. Terry Lowry, *The Battle of Scary Creek*, (Pictorial Histories Publishing Co., Chas., WV., 1982) p. 162

3. United States War Department, *War of the Rebellion: A Compilation of the Official Records of the Union and Confederate Armies*, 70 vols. in 128 books, (Washington: Government Printing Office, 1881-1901), Series 1, V. 2 p. 995

4. Diary of Victoria Hansford Teays, transcript courtesy of Bill Wintz, Upper Vandalia Historical Society, P. O. Box 517, Poca, WV., 25159

5. George E. Moore, *A Banner in the Hills*, (New York: Appleton-Century-Crofts, Meredith Publishing Co., 1963.) p. 103

6. Official Records, V. 51 pt. 1 p. 426

7. Albert D. Richardson, *The Secret Service, The Field, The Dungeon, and The Escape*, (American Publishing Company, Hartford, Conn., 1865) p. 174

8. James H. Mays, Lee May, ed., *Four Years for Old Virginia*, (Privately printed) The war time experiences of James H. Mays, Company F, 22nd Virginia Infantry, assembled by his son.

9. *West Virginia Review*, magazine, issue of October 1925.

10. William Bahlmann, "Down in the Ranks," *Journal of the Greenbrier Historical Society*, Oct. 1970, V. 2 No. 2 p. 51

11. Official Records, V. 5 p. 770

12. Beuhring H. Jones, "My First Thirty Days Experience as a Captain." *Southern Literary Messenger*, V. 37 No. 2, 1863

13. Diary of A. B. Roler, Wise Legion, July-Sept. 1861, in the manuscript collections of the Virginia Historical Society, Richmond, Va.

14. Stutler, p. 76

15. B. H. Jones, "My First Thirty Days Experience as a Captain."

16. Official Records, V. 2 p. 1012

17. Jacob Dolson Cox, *Military Reminiscenses*, (New York: Charles Scribner & Sons, 1900) V. 1, p. 78

18. Ibid., p. 81

19. Ambrose Bierce, *The Collected Works of Ambrose Bierce*, (New York, Gordian Press 1906-1966), V. 1, p. 227.

20. Cox, p. 83

21. Col. Charles Whittlesey, *War Memoranda: Cheat River to the Tennessee 1861-62*, (Cleveland, Ohio, 1884) p. 26

22. Official Records, V. 51 pt. 1, p. 439

23. Cox, p. 90

NOTES TO CHAPTER 3

1. United States War Department, *War of the Rebellion, A Compilation of the Official Records of the Union and Confederate Armies*, 70 vols. in 128 books, (Washington: Government Printing Office, 1881-1901), Series 1, V. 5, p. 773

2. B. Estvan, *War Pictures From The South*, (New York: D. Appleton & Co., 1863), p. 117

3. Henry Heth, (James I. Robertson, Jr. editor), "Memoirs of Henry Heth," *Civil War History*, V. 8 No. 1, p. 13

4. Official Records, V. 5 p. 773

5. Ibid., p. 774

6. Official Records, V. 51 pt. 2 p. 224

7. Official Records, V. 5 p. 778

8. Ibid., V. 5 p. 774

9. Ibid., V. 5 p. 781

10. Ibid., V. 5 p. 784

11. Ibid., V. 5 p. 785

12. Ibid., V. 5 p. 785

13. Ibid., V. 5 p. 789

14. Joshua Horton & Solomon Teverbaugh, *History of the 11th O.V.I.*, (Dayton, Ohio, W. J. Shuey, 1866), p. 38

15. Official Records, V. 51 pt. 1, p. 448

16. "The Old Stone House," *West Virginia History Magazine*, January 1971, p. 115

17. Official Records, V. 5 p. 798

18. Lawrence Wilson, *Itinerary of the 7th Ohio Volunteer Infantry 1861-64*, (New York & Washington: Neale Publishing Co., 1907), p. 61

19. Horton & Teverbaugh, p. 41

20. "The Bushwackers' War," *Civil War History*, V. 10 p. 421

21. Official Records, V. 5 p. 799

22. Wilson, p. 69

23. Horton & Teverbaugh, p. 216

24. Official Records, V. 5 p. 156

25. Official Records, V. 51 pt. 2, p. 248

26. Official Records, V. 5 p. 804

27. Ibid., p. 810

28. Ibid., p. 115

29. Jacob Dolson Cox, *Military Reminiscenses*, (New York: Charles Scribner & Sons, 1900), V. 1, p. 98

30. Official Records, V. 5 p. 815

31. Ibid., p. 812

32. "Civil War Letters From The Kanawha Valley," *West Virginia Heritage*, V. 2 No. 41, October 14, 1967. From original letters in the manuscript collections of the University of North Carolina Library.

33. Terry Lowry, *September Blood, The Battle of Carnifex Ferry*, (Pictorial Histories Publishing Co., Chas., WV. 1985), p. 28

34. Official Records, V. 5 p. 117

35. National Archives, Confederate Records Group 109, CSR, Records of R. A. Bailey, 22nd Va. Inf., Letter of August 28, 1861
36. Record Book of Jacob D. Cox, May-November 1861, Oberlin College, Oberlin, Ohio. p.201
37. Cox, *Military Reminiscenses*, p. 102.
38. Official Records, V. 51 pt. 2, p. 266
39. Ibid., p. 267
40. Official Records, V. 5 p. 123
41. Horton & Teverbaugh, p. 42.
42. Official Records, V. 5 p. 123
43. From the Record Book of Alfred Beckley, 27th Brigade Va. Militia, August-October, 1861, entry of September 4, 1861. Microfilm copy in the West Virginia State Archives, Charleston, WV.
44. Frank Moore, *The Rebellion Record*, (New York, G. P. Putnam, 1861-71), V. 4, p. 31, poetry section.
45. Official Records, V. 51 pt. 2, p. 275
46. Record Book of Alfred Beckley, entry of September 4, 1861.
47. Official Records, V. 51 pt. 2, p. 270
48. Ibid., pt. 1, p. 473
49. Ibid., p. 473
50. "September Blood," Appendix.
51. Official Records, V. 5 p. 146
52. Official Records, V. 51 pt. 2, p. 296
53. Ibid., p. 298 — Also Beckley Record Book, entry of September 17, 1861.
54. Henry Heth, p. 15
55. Official Records, V. 51 pt. 1, p. 481.
56. Ibid., p. 481

NOTES TO CHAPTER 4

1. From a letter published in the *Richmond Enquirer*, October 1, 1861.
2. United States War Department, *War of the Rebellion: A Compilation of the Official Records of the Union and Confederate Armies*, 70 vols. in 128 books, (Washington: Government Printing Office, 1881-1901), V. 5 p. 859
3. Ibid., p. 860
4. Ibid., p. 864
5. Official Records, V. 51 pt. 2, p. 303
6. Official Records, V. 5 p. 868
7. Ibid., p. 868
8. Ibid., p. 162
9. Ibid., p. 874
10. Ibid., p. 874
11. From a letter published in the *Richmond Enquirer*, October 1, 1861.
12. Official Records, V. 51 pt. 1, p. 487
13. Ibid., p. 486
14. "A Virginian's Dilemma," The Civil War Diary of Isaac Noyes Smith, 22nd Va. Inf., September-November 1861. As printed in *West Virginia History*, April 1966, p. 184.

15. G. Moxley Sorrel, *Recollections of a Confederate Staff Officer*, (New York, 1905), p. 75

16. "A Virginian's Dilemma," p. 185

17. Ibid., p. 185

18. Walter H. Taylor, *Four Years With General Lee*, (Indiana University Press, 1962), p. 33.

19. T. C. Morton, "Recollections of General Lee," as published in the *Southern Historical Society Papers*, V. 11 p. 519.

20. Douglas S. Freeman, *R. E. Lee*, (New York, 1934-35), V1 p. 591

21. Burke Davis, *Gray Fox, Robert E. Lee and the Civil War*, (New York, The Fairfax Press, 1956), p. 50

22. Official Records, V. 51 pt. 1, p. 486

23. Ibid., p. 487

24. Ibid., p. 488

25. Joshua Horton & Solomon Teverbaugh, *History of the 11th O.V.I.*, (Dayton, Ohio, W. J. Shuey, 1866), p. 46

26. Official Records, V. 5 p. 879

27. Official Records, V. 51 pt. 2, p. 313

28. Freeman, p. 593

29. Jacob Dolson Cox, *Military Reminiscenses*, (New York, Charles Scribner & Sons, 1900), p. 119

30. Henry Heth, (James I. Robertson, Jr. editor), "Memoirs of Henry Heth," *Civil War History*, V. 8 No. 1, p. 16

31. Charles T. Quintard, *Dr. Quintard, Chaplain C.S.A. and Second Bishop of Tennessee*, (A. H. Noll, editor, Sewanee, Tenn., 1905) p. 32

32. "A Virginian's Dilemma," p. 186.

33. J. M. Miller, *Recollections of a Pine Knot, Campaigns of West Virginia, Kentucky and Fort Donelson*, (Commonwealth Publishing Co., Greenwood, Mississippi, 1899), p. 7

34. From the Papers of Hugh B. Ewing, 30th O.V.I., in the manuscript collections of the Ohio Historical Society, Columbus, Ohio, letter from Hugh B. Ewing to his wife, October 31, 1861.

35. Official Records, V. 51 pt. 2, p. 320

36. Cox, p. 121

37. Taylor, p. 35

38. "Memoirs of Leroy W. Cox, Experiences of a Young Soldier in the Confederacy," unpublished personal narrative, Manuscripts Collection, Virginia Historical Soc.

39. Numerous historians have recorded that Loring brought 9,000 men to the Sewell Mountain area. Actually, be brought approximately 3,000 troops, the balance of his command remaining in the Tygart Valley area. It has been variously reported that Lee had from 12,000 to 25,000 troops at Sewell. His command at no time exceeded 9,000 men fit for duty.

40. Walter Womack, ed., *The Civil War Diary of Capt. J. J. Womack, Co. E, 16th Tennessee Volunteers*, (McMinnville, Tenn., Womack Printing Co., 1961), p. 19

41. From an unsigned letter to the editor of the *Nashville Union and American*, October 11, 1861. Original in the manuscript collections of the Tennessee State Archives, Nashville, Tenn.
42. Official Records, V. 51 pt. 2, p. 324
43. Cox, p. 120
44. Ibid., p. 120
45. The majority of the 2,000 men Floyd brought up from Meadow Bluff were the Militia forces of General Chapman who were released from duty two weeks after their arrival at Sewell.
46. "A Virginian's Dilemma," p. 188
47. Womack, p. 20
48. From the unpublished diary of Samuel J. Mullins, Capt., Co. A, 42nd Va. Infantry. Copies provided courtesy of Mr. R. P. Gravely, Martinsville, Va.
49. "A Virginian's Dilemma," p. 188
50. Cox, p. 122-24
51. "A Virginian's Dilemma," p. 190
52. From an unsigned letter to the editor of the *Nashville Union and American*, October 21, 1861.
53. National Archives, Records of the Adjutant Generals Office, Record Group 94, Microcopy Publication M 1098, Roll # 6, The post war papers of General Henry Benham, 1873.
54. Official Records, V. 51 pt. 2, p. 335
55. Ibid., p. 335
56. Heth, p. 19
57. R. E. Lee, Jr., *Recollections and Letters of General Robert E. Lee*, (New York, 1904), Lee's letter of October 7, 1861.
58. Record Book of Jacob D. Cox, May-November 1861, Oberlin College, Oberlin, Ohio, p. 348.
59. Ibid., p. 352
60. Official Records, V. 51 pt. 1, p. 494-95
61. Ellen W. Tompkins (ed.), "The Colonel's Lady: Some Letters of Ellen W. Tompkins: July-December 1861," *Virginia Magazine of History and Biography*, Oct., 1960 #69, p. 409.
62. Cox, p. 126
63. Official Records, V.51 pt. 2, p. 341-42
64. From a letter written by the editor of the *Lynchburg Republican*, October 12, 1861. As published in the *Nashville Union and American*, October 24, 1861.
65. From the unpublished diary of Carrol H. Clark, Co. I, 16th Tennessee Inf., in the manuscript collections of the Tennessee State Archives, Nashville, Tenn.
66. From an unpublished letter in the Wartime Papers of Robert E. Lee, 1861, Manuscript Collections, Virginia Historical Society, Richmond, Va.
67. Ibid., Report of Bedford Broun, Surgeon, 14th North Carolina Infantry.
68. Official Records, V. 51 pt. 2, p. 347
69. Official Records, V. 5 p. 900-02.
70. Mullins Diary, 42nd Va. Inf., October 1861.

71. From the CSR Roster, as printed in the regimental history, 42nd Va., Inf., by John Chapla, (H. E. Howard Publishing Co., Lynchburg, Va., 1983).
72. Official Records, V. 5 p. 908-09.
73. From an unsigned letter to the editor of the Richmond Enquirer, October 31, 1861. As printed in the Nashville Union and American, November 5, 1861.
74. From an unsigned letter to the editor of the Richmond Dispatch, November 1, 1861. As printed in the Nashville Union and American, November 5, 1861.
75. Official Records, V. 5 p. 917-18.
76. Cox, p. 126
77. Official Records, V. 5 p. 924.
78. Official Records, V. 51 pt. 2, p. 360.
79. Ibid., p. 361-62.
80. Taylor, p. 34-35.
81. A. L. Long, Memoirs of Robert E. Lee, (New York, 1886), p. 493-94.

NOTES TO CHAPTER 5

1. "A Virginian's Dilemma," The Civil War Diary of Isaac Noyes Smith, 22nd Va. Inf., September-November 1861. As printed in West Virginia History, April 1966, p. 193.
2. William Forse Scott, Philander P. Lane, Colonel of Volunteers in the Civil War, 11th Ohio Infantry, (Privately printed 1920), p. 57.
3. Frank Moore, The Rebellion Record, (New York, G. P. Putnam, 1861-71), V. 2, Document 136, p. 301.
4. From research papers of the Trans Allegheny Historical Association, Box 65, Beaver, WV., 25813, Mr. Jody Mays.
5. Scott, p. 57.
6. United States War Department, War of the Rebellion, A Compilation of the Official Records of the Union and Confederate Armies, 70 vols. in 128 books, (Washington: Government Printing Office, 1881-1901), Series 1, V. 5 p. 286.
7. Scott, p. 57.
8. Frank Moore, V. 2, Doc. 136 p. 301.
9. Papers of Frank Jones, 13th OVI., Ohio Historical Society, Columbus, Ohio.
10. Scott, p. 58.
11. "D. B. Baldwin in the Skirmish at Gauley Bridge," West Virginia History, V. 24, p. 352.
12. From unknown newspaper clippings in the files of the Fayette County Historical Society, Ansted, WV.
13. Rutherford B. Hayes, (ed., Charles R. Williams), Diary and Letters of Rutherford B. Hayes, (The Ohio State Archaeological and Historical Soc. 1922), V. 2 p. 135.
14. Frank Jones letters, OHS
15. Jacob Dolson Cox, Military Reminiscenses, (New York: Charles Scribner & Sons, 1900), V. 1, p. 134.
16. Scott, p. 57.

17. Letters of Frank Jones, OHS.
18. Hayes, V. 2 p. 136.
19. Scott, p. 58.
20. Joshua Horton & Solomon Teverbaugh, *History of the 11th O.V.I.*, (Dayton, Ohio, W. J. Shuey, 1866), p.253.
21. Frank Moore, V. 2 p. 303
22. Henry Heth, (James I. Robertson, Jr. editor), "Memoirs of Henry heth," *Civil War History*, V. 8 No. 1, p. 17.
23. From the *Fayette Tribune*, January 8, 1933.
24. Letters from the 45th Virginia Infantry in the manuscript collections of the West Virginia University, Morgantown, WV.
25. Official Records, V. 5 p. 286.
26. From the unpublished papers of John W. Hodnett, 13th Georgia Infantry, in the manuscript collections of the William R. Perkins Library, Duke University, Durham, N. C., 27706.
27. National Archives, Records of the Adjutant Generals Office, Record Group 94, Microcopy Publication M-1098, Roll # 6, The post war papers of General Henry Benham, 1873.

NOTES TO CHAPTER 6

1. United States War Department, *War of the Rebellion: A Compilation of the Official Records of the Union and Confederate Armies*, 70 Vols. in 128 books, (Washington: Government Printing Office, 1881-1901), Series 1, V. 5. p. 947.
2. From the unpublished letters of Isaac Noyes Smith in the manuscript collections of the Virginia Historical Society, Richmond, VA.
3. Frank Moore, *The Rebellion Record*, (New York, G. P. Putnam, 1861-71), V. 2 p. 353.
4. Official Records, V. 5 p. 272-73.
5. Official Records, V. 51 pt. 2 p. 375.
6. Henry Heth, (James I. Robertson, Jr. editor), "Memoirs of Henry Heth," *Civil War History*, V. 8 no. 1 p. 18.
7. Official Records, V. 5 p. 283.
8. Frank Moore, V. 2 p. 388.
9. National Archives, Records of the Adjutant Generals Office, Record Group 94, Microcopy Publication M-1098, roll #6, The post war papers of General Henry Benham, 1873.
10. Official Records, V. 5 p. 257.
11. Benham papers, 1873.
12. Frank Moore, V. 2 p. 387.
13. Ibid., p. 383
14. B. Estvan, *War Pictures From the South*, (New York: D. Appleton & Co., 1863), p. 130.
15. Frank Moore, V. 2 p. 383.
16. Benham papers, 1873.
17. Frank Moore, V. 2 p. 389
18. Ibid., p. 383.

19. Official Records, V. 5 p. 284.

20. "A Virginian's Dilemma," The Civil War Diary of Isaac Noyes Smith, 22nd Va. Inf., September-November 1861. As printed in *West Virginia History*, April 1966, p. 198.

21. Official Records, V. 5 p. 955.

22. Ibid., p. 656.

23. Ibid., p. 670.

24. Ellen W. Tompkins (ed.) "The Colonel's Lady: Some Letters of Ellen W. Tompkins, July-December 1861," *Virginia Magazine of History and Biography*, Oct. 1960 #69 p. 415-17.

25. Official Records, V. 5 p. 470.

26. Rutherford B. Hayes, (The Ohio State Archaeological and Historical Soc. 1922), *Diary and Letters of Rutherford B. Hayes*, V. 2 p. 164-170.

27. Official Records, V. 51 pt. 2 p. 429.

NOTES TO CHAPTER 7

1. "The Bushwackers' War," *Civil War History*, V. 10 p. 418.

2. United States War Department, *War of the Rebellion: A Compilation of the Official Records of the Union and Confederate Armies*, 70 vols. in 128 books, (Washington: Government Printing Office, 1881-1901), Series 1, V. 12 pt. 3, p. 11.

3. Ibid., p. 46.

4. "The Destruction of Gauley Bridge," *West Virginia Review*, October 1925.

5. Official Records, V. 12 pt. 3 p. 47.

6. Official Records, V. 51 pt. 2 p. 517.

7. Official Records, Series 2, V. 2 – Prisoner of War lists. Compiled from general listings of prisoners held by various military departments, March 1862.

8. From the unpublished letters of Albert George, 30th OVI -in the manuscript collections of the Ohio Historical Soc., Columbus, Ohio.

9. From the letters of William Ludwig, in the manuscript collections of the West Virginia University, Morgantown, WV.

10. William W. Lyle, *Lights and Shadows of Army Life*, Reverend W. W. Lyle, 2nd. edition, (Cincinnati, R. W. Carroll & Co., 1865), p. 41

11. From originals in the possession of Mr. A. N. Gwinn, Grand Rapids, Mich.

12. From papers in the Roy Bird Cook Collection, WVU, Morgantown, WV.

NOTES TO CHAPTER 8

1. From a letter by George W. Botkin found in the Baker papers, VFM 1727, Manuscript Collections of the Ohio Historical Society, Columbus, Ohio.

2. George E. Moore, *A Banner in the Hills*, (New York: Appleton-Century-Crofts, Meredith Publishing, 1963), p. 167.

3. United States War Department, *War of the Rebellion: A Compilation of the Official Records of the Union and Confederate Armies*. 70 vols. in 128 books, (Washington: Government Printing Office, 1881-1901), V. 19 pt. 1 p. 1058-59.

4. Jack L. Dickinson, *8th Virginia Cavalry*, (H. E. Howard Publishing Co., Lynchburg, Va., 1986), p. 33

5. Official Records, V. 19 pt. 1 p. 1059.

6. Dickinson, p. 34.

7. From the unpublished diary of Samuel Harrison, 44th OVI—in the manuscript collections of the Ohio Historical Society, Columbus, Ohio.

8. Official Records, V. 19 pt. 1 p. 1085, Report of Col. Browne, 45th Va. Inf.

9. Ibid., p. 1088.

10. James A Davis, *51st Virginia Infantry*, (H. E. Howard Publishing Co., Lynchburg, Va., 1984), p. 15

11. Montgomery, *The Ninth Reunion of the 37th OVI—September 10-11, 1889*, (Privately Printed, Toledo, Ohio, 1889), p. 10.

12. Officials Records, V. 19 pt. 1 p. 1082.

13. "My Recollections of the War," by F. G. Shackelford, as published in the *Nicholas Chronicle*, Summersville, WV., 1895.

14. Official Records, V. 19 pt. 1 p. 1059.

15. Harrison Diary, Ohio Historical Society.

16. Official Records, V. 19 pt. 1 p. 1062.

17. C. Shirley Donnelly, *Historical Notes on Fayette County*, WV., (Privately Printed, 1958), Speech of John L. Vance, Lt. Col. 4th WV. Inf., 1896.

18. Official Records, V. 19 pt. 1 p. 1079.

19. Frank Moore, *The Rebellion Record*, (New York, G. P. Putnam, 1861-71), V. 5, Report of Colonel Toland.

20. Official Records, V. 19 pt. 1 p. 1062.

21. Shackelford, Recollections.

22. Harrison Diary, Ohio Historical Society.

23. J. T. Peters & H. B. Carden, *History of Fayette County, WV.*, (Jarrett Printing Company, Chas., WV. 1926) Quoting Cpt. Joel H. Abbott, p. 217.

24. Official Records, V. 19 pt. 1 p. 1083

25. Davis, p. 16.

26. Harrison Diary, Ohio Historical Society.

27. Diary of Virtoria Hansford Teays, transcript courtesy of Bill Wintz, Upper Vandalia Historical Society, P. O. Box 517, Poca, WV. 25159.

28. Official Records, V. 19 pt. 1 p. 1072-73.

29. Oren F. Morton, *A History of Monroe County West Virginia*, (Regional Publishing Co., Baltimore, 1974), containing the Civil War Diary of the Reverend S. R. Houston, p. 166-179., entry of September 17, 1862, quoted.

30. Tom Taylor papers, 47th O.V.I., Ohio Historical Society.

31. Official Records, V. 19 pt. 2 p. 615.

32. Ibid., p. 625.

33. Ibid., p. 627.

34. Ibid., p. 637.

35. Ibid., p. 656.

36. Ibid., p. 661.

37. Ibid., p. 684.

38. Ibid., p. 689.

39. Ibid., p. 691.

NOTES TO CHAPTER 9

1. United States War Department, *War of the Rebellion: A Compilation of the Official Records of the Union and Confederate Armies*, 70 vols. in 128 books, (Washington: Government Printing Office, 1881-1901), V. 19 pt. 2 p. 537.
2. Ibid., p. 556-57.
3. From the letters of George B. Turner, 91st O.V.I.—in the manuscript collections of the Ohio Historical Society, Columbus, Ohio.
4. J.E.D. Ward, *12th Ohio Volunteer Infantry*, (Ripley, Ohio, 1864), p. 61.
5. "War Times in Mountain Cove," The letters of Nancy Hunt, 1862-65, in the manuscript collections of the West Virginia University Library, Morgantown, WV. Copies provided courtesy of Dr. Otis K. Rice, West Virginia Tech, Dept. of History, Montgomery, WV.
6. T. Harry Williams, *Hayes of the Twenty Third*, (New York: Alfred A. Knopf, 1965), p. 144.
7. Turner letters, Ohio Historical Society.
8. Williams, p. 146.
9. Official Records, V. 21 p. 992-93.
10. Williams, p. 147.
11. National Archives, Returns from U. S. Military Posts, 1800-1916, Microcopy No. 617, Roll #1511, containing post returns for Gauley Bridge, WV., and others.

NOTES TO CHAPTER 10

1. Milton W. Humphreys, *Military Operations, 1861-63, Fayetteville, WV.*, (Privately printed at Fayetteville, WV., 1926), p. 22.
2. Ibid., p. 22.
3. From the unpublished letters of W. S. Newton, Surgeon, 91st OVI—in the manuscript collections of the Ohio Historical Society, Columbus, Ohio.
4. From the Civil War letters of Gordon Thompson, 60th Va. Infantry. Copies provided courtesy of Roger Thompson, 130 W. 11th Ave., Huntington, WV.
5. Frank Moore, *The Rebellion Record*, (New York, G. P. Putnam, 1861-71), V. 6 Document 195.
6. Joseph Alliene Brown, *The Memoirs of a Confederate Soldier*, (Copyright 1940, by Sam Austin, The Forum Press, Abingdon, Va.), p. 6.
7. Newton letters, Ohio Historical Society.
8. John L. Scott, *36th Virginia Infantry*, (H. E. Howard Publishing Co., Lynchburg, Va. 1987), p. 20.
9. From the Civil War letters of John V. Young, in the manuscript collections of the West Virginia University Library, Morgantown, WV. Roy Bird Cook papers.
10. Newton letters, Ohio Historical Society.
11. J. T. Peters & H. B. Carden, *History of Fayette County WV.*, (Jarrett Printing Company, Chas., WV. 1926), p. 218, quoting Cpt. Joel H. Abbott.
12. Whitelaw Reid, *Ohio in the War*, (Two volumes, Cinn., New York: Moore-Wilstach and Baldwin, 1868), V. 1 p. 799-800.

13. A. H. Windsor, *History of the 91st OVI* – (Cinn., Gazette Steam Printing House, 1865), p. 40.

14. From unidentified newspaper clippings in the files of the Fayette County Historical Society, Ansted, WV.

15. "War Times in Mountain Cove," The Letters of Nancy Hunt, 1862-65, in the manuscript collections of the WVU Library, Morgantown, WV.

16. National Archives, Returns From U. S. Military Posts, 1800-1916, Microcopy No. 617 Roll #1510, containing returns for Fayetteville, WV., and other locations.

17. Peters & Carden, p. 218.

Bibliography

■ Books

Bierce, Ambrose. *The Collected Works of Ambrose Bierce*. New York: Gordian press 1906-1966.

Brown, Joseph A. *The Memoirs of a Confederate Soldier*. Abingdon, Va: The Forum Press, 1940.

Brown, William G. *History of Nicholas County W.VA*. Richmond: Dietz Press, 1954.

Chapla, John. *42nd Virginia Infantry*. Lynchburg, Va: H. E. Howard Pub. Co., 1983.

Civil War Centennial Commission of Tennessee. *Tennesseans in the Civil War*. 2 vols. Nashville, Tenn. 1964 – reprinted 1985.

Cohen, Stan. *A Pictorial History of the Civil War in West Virginia.*: 1975. Charleston, WV. Pictorial Histories Pub. Co.

Cox, Jacob D. *Military Reminiscenses*. New York: Charles Scribner & Sons, 1900.

Cutchins, John A. *A Famous Command: The Richmond Light Infantry Blues*. Richmond: Garrett & Massie, 1934.

Davis, Burke. *Gray Fox: Robert E. Lee and the Civil War*. New York: Fairfax Press, 1956.

Davis, James A. *51st Virginia Infantry*. Lynchburg, Va: H. E. Howard Pub. Co., 1984.

Darlington, L. Neil. *Cabins of the Loop & Environs of Fayette Co., WV*. Parsons, WV: McClain Printing Co., 1988.

Dickinson, Jack L. *8th Virginia Cavalry*. Lynchburg, Va: H. E. Howard Pub. Co., 1986.

Donnelly, C. Shirley. *Historical Notes on Fayette Co. WV*. Privately printed, 1958.

Egan, Michael. *The Flying Gray-Haired Yank*. Philadelphia: Hubbard Brothers, 1888.

Estvan, B. *War Pictures From the South*. New York: D. Appleton & Co., 1863.

Evans, Clement A. ed., *Confederate Military History*. 12 Volumes. Atlanta: Confederate Pub. Co., 1899.

Faust, Patricia. ed., *Historical Times Illustrated Encyclopedia of the Civil War*. New York: Harper & Row, 1986.
Freeman, Douglas S. *R. E. Lee*. 2 Volumes. New York, l934-35.
Hayes, Rutherford B. and Charles R. Williams, ed., *Diary and Letters of Rutherford B. Hayes*. The Ohio State Archaeological and Historical Soc. 1922.
Head, Thomas A. *Campaigns and Battles of the 16th Regiment Tennessee Volunteers*. Nashville: Cumberland Presbyterian Pub. House, 1885.
Horton, Joshua & Solomon Teverbaugh. *History of the 11th OVI*. Dayton, Ohio: W. J. Shuey, 1866.
Howe, Henry. *Historical Collections of Ohio*. Columbus: Howe & Son, 1890-91.
Humphreys, Milton W. *Military Operations 1861-1863 at Fayetteville, WV*. Privately printed, 1926.
Kempfer, Lester L. *The Salem Light Guard—36th OVI*. Chicago: Adams Press, 1973.
Lamers, William A. *The Edge of Glory: A Biography of Gen. William S. Rosecrans*. New York: Harcourt—Brace & World, 1961.
Lang, Theodore F. *Loyal West Virginia From 1861-65*. Baltimore: Deutch Pub., 1895.
Lee, Robert E. Jr. *Recollections and Letters of Gen. Robert E. Lee*. New York: 1904.
Long, A. L. *Memoirs of Robert E. Lee*. New York: 1886
Lowry, Terry. *The Battle of Scary Creek*. Charleston, WV: Pictorial History Pub. Co., 1982.
Lowry, Terry. *September Blood: The Battle of Carnifex Ferry*. Charleston, WV: Pictorial Histories Pub. Co., 1985.
Lyle, W. W. *Lights and Shadows of Army Life—11th OVI*. Cincinnati: Carroll & Co., 1865.
Mays, James H. and Lee Mays, ed., *Four Years For Old Virginia*. Privately printed, 1972.
Miller, J. M. *Recollections of a Pine Knot: Campaigns of West Virginia, Kentucky and Fort Donelson*. Greenwood, Mississippi: Commonwealth Pub. Co., 1899.
Montgomery, _____. *The Ninth Reunion of the 37th OVI*. Toledo, Ohio: Privately printed, 1889.
Moore, Frank. *The Rebellion Record*. New York: G. P. Putnam, 1861-71.
Moore, George E. *A Banner in the Hills*. New York: Appleton -Century-Crofts, 1963.
Morton, Oren F. *A History of Monroe County WVA*. Baltimore: Regional Pub. Co., 1974.
Peters, J. T. & H. B. Carden. *History of Fayette Co. WVA*. Charleston, WV: Jarrett Printing Co., 1926.
Quintard, Charles T. *Dr. Quintard: Chaplain C.S.A. and Second Bishop of Tennessee*. Sewanee, Tenn: A. H. Noll, 1905.
Reid, Whitelaw. *Ohio In The War*. Cincinnati: New York: Moore & Baldwin, 1868.
Rice, Otis K. *A History of Greenbrier County WVA*. Parsons, WV: McClain Printing Co., 1986.

Richardson, Albert D. *The Secret Service: The Field, The Dungeon, and The Escape*. Hartford, Conn: American Pub. Co., 1865.

Scott, John L. *36th Virginia Infantry*. Lynchburg, Va: H. E. Howard Pub. Co., 1987.

Scott, William F. *Philander P. Lane: Colonel of Volunteers In The Civil War, 11th Ohio Infantry*. Privately printed, 1920.

Shetler, Charles. *West Virginia Civil War Literature*. Morgantown, WV: WVU Press, 1963.

Sorrell, G. M. *Recollections of a Confederate Staff Officer*. New York: 1905.

Strother, David H. *A Virginia Yankee In The Civil War: The Diaries of David Hunter Strother*. Chapel Hill: UNC Press, 1961.

Stutler, Boyd. *The Civil War in West Virginia*. Charleston, WV: Education Foundation, 1963.

Sutton, J. J. *History of the Second Regiment W.Va. Cavalry Volunteers*. Portsmouth, Ohio: 1892.

Taylor, Walter H. Dr. James I. Robertson Jr. ed., *Four Years With General Lee*. Indiana University Press, reprint, 1962.

United States War Department, *War of the Rebellion: A Compilation of the Official Records of the Union & Confederate Armies*. Washington: Government Printing Office, 1881-1901, 70 vols. in 128 books.

Wallace, Lee A. *A Guide to Virginia Military Organizations 1861-65*. Richmond: Virginia Civil War Commission, 1964.

Ward, J. E. D. *12th Ohio Volunteer Infantry*. Ripley Ohio: 1864.

Warner, Ezra. *Generals in Blue*. Baton Rouge: LSU Press, 1964.

Warner, Ezra. *Generals in Gray*. Baton Rouge: LSU Press, 1959.

Whittlesey, Charles. *War Memoranda: Cheat River to the Tennessee 1861-62*. Cleveland, Ohio: 1884.

Williams, T. H. *Hayes of the Twenty Third*. New York: Alfred A Knopf, 1965.

Wilson, Lawrence. *Itinerary of the 7th Ohio Volunteer Infantry 1861-64*. New York & Washington: Neale Pub. Co., 1907.

Windsor, A. H. *History of the 91st OVI*. Cincinnati: Gazette Steam Pub. Co., 1865.

Wise, John S. *End of an Era*. New York: A. S. Barnes & Co., 1965.

Womack, Walter. ed., *The Civil War Diary of Capt. J. J. Womack, Co. E. 16th Tennessee Volunteers*. McMinnville, Tenn: Womack Printing Co., 1961.

Zinn, Jack. *R. E. Lee's Cheat Mountain Campaign*. Parsons, WV: McClain Printing Co., 1975.

■ Magazines & Journals

Civil War History. "The Memoirs of Henry Heth," Vol. 8 No. 1 p. 13.

Civil War History. "The Bushwackers' War," Vol. 10 p. 421.

Journal of the Greenbrier Historical Society. William Bahlman, "Down In The Ranks," October 1970, Vol. 2 No. 2 p. 43.

Southern Literary Messenger. Beuhring H. Jones, "My First Thirty Days Experience as a Captain," Vol. 37 No. 2, 1863. The Vandalia Journal. William Wintz, "The Great Kanawha Valley Flood," Jan. 1984.

Virginia Magazine of History & Biography. "The Colonel's Lady: Some Letters of Ellen W. Tompkins, July-December, 1861," October 1960 No. 69 p. 387-419.

West Virginia History Magazine, (WVH) in order of date:

WVH, July 1944, "The Campaigns of McClellan & Rosecrans in West Virginia."

WVH, October 1944, "Jacob Dolson Cox in West Virginia."

WVH, April 1947, "Gen. John B. Floyd and the West Virginia Campaigns of 1861."

WVH, July 1953, "Fayetteville, West Virginia During the Civil War."

WVH, January 1961, "The Romance of a Man in Grey," including the love letters of Captain James S. Peery, 45th Va. Infantry.

WVH, January 1962, "The Civil War Comes to Charleston."

WVH, July 1963, "Lincoln and West Virginia Statehood."

WVH, April 1964, "The Presidential Election of 1860 in West Virginia."

WVH, April 1965, "Colonel George S. Patton and the 22nd Va. Infantry."

WVH, April 1966, "A Virginian's Dilemma," The Civil War Diary of Isaac Noyes Smith, 22nd Va. Infantry, September-November, 1861.

WVH, October 1969, "The Unfortunate Military Career of Henry A. Wise in West Va."

WVH, January 1971, "The Old Stone House."

WVH, January 1973, "Major Cunningham's Journal of 1862."

WVH, April 1982, "The Civil War Letters of Laban Gwinn."

WVH, Volume 24, p. 352, "D.B. Baldwin in the Skirmish at Gauley Bridge."

West Virginia Review, (WVR) in order of date:

WVR, October 1925, "The Destruction of Gauley Bridge."

WVR, November 1930, "The Journal of a Soldier of 1861."

WVR, December 1933, "Joseph A.J. Lightburn, the Fighting Parson."

WVR, May 1934, "Albert Gallatin Jenkins, a Confederate Portrait."

WVR, September 1934, "Episode at Big Sewell."

WVR, October 1935, "After the Battle of Carnifex Ferry."

WVR, March 1942, "Charleston's Dunkirk."

WVR, October 1946, "The Death of Colonel St. George Croghan."

■ Newspapers & Newspaper Articles

Charleston WV. Gazette, Sept. 17, l922. "Bahlman Gives History of Men Serving in '61."

Charleston Gazette, July 12, 1925. "D. C. Gallagher on Civil War in the Valley."

Charleston Gazette, Jan. 10, 1926. "Civil War in the Kanawha Valley Saw Many Engagements."

Charleston Daily Mail, Feb. 15, 1963. "Rag-Tag Southern Soldiery Swamped Charleston in '61."

Fayette Tribune, Fayetteville, WV. Jan. 8, 1933. "Recollections of A. W. Hamilton."

The Guerilla, Charleston, WV. Issues of September 1862, published by the Confederate forces occupying the Kanawha Valley in 1862. Microfilm copies at the W. Va. State Archives, Charleston.

Kanawha Valley Star, June 4, 1860–July 2, 1861. Microfilm collection, W.Va. State Archives, Charleston.
Nashville Union And American, Nashville, Tennessee. Various issues, August–November 1861.
New York Herald (NYH) and New York Times (NYT), in order of date:
NYH, August 2, 1861. "The Occupation of Charleston, Va."
NYH, August 23, 1861. "Skirmish At The Hawks Nest."
NYH, August 29, 1861. "Important News From The Kanawha Valley."
NYH, September 17, 1861. "Retreat of The Rebels Wise & Floyd."
NYH, October 4, 1861. "Reported Battle Between General Cox & The Rebels."
NYH, October 7, 1861. "The Seat Of War In Western Virginia."
NYH, October 30, 1861. "Important News From Western Virginia."
NYH, November 8, 1861. "The Fighting At Gauley Bridge & Cotton Hill."
NYT, November 10, 1861. "The Fighting Between Floyd & Rosecrans."
NYT, November 23, 1861. "The Scene of Floyds Repulse."
Nicholas County Chronicle, Nicholas Co. W.Va., March–November 1895. "The Recollections of F. G. Shackelford, 36th Va. Infantry."
Richmond Va. Dispatch, various issues between July and December 1861.
Richmond Va. Enquirer, various issues between August and November 1861.

■ Manuscripts & Narratives

W. Baker papers, 1st Kentucky Infantry, Ohio Historical Society, Columbus, Oh.
Brigade record book of Alfred Beckley, 27th Brigade Va. Militia. W.Va. Archives.
Elijah Beeman Letters, 12th O.V.I., Cabell County Public Library, Huntington, WV.
Post war papers of General Henry Benham, National Archives, Records of the Adjutant Generals Office, RG 94, Microcopy M-1098, Roll #6.
Diary of Carroll Clark, 16th Tennessee Infantry. Manuscript collections of the Tennessee State Archives, Nashville.
Roy Bird Cook Collection, various items relating to the war in W.Va. Manuscript collections of the West Virginia University, Morgantown, WV.
Regimental record book of General Jacob D. Cox, May-November 1861. Oberlin College, Oberlin, Ohio.
Memoirs of Leroy Wesley Cox: Experiences of a Young Soldier in the Confederacy. Unpublished personal narrative, Manuscripts Collection, Va. Historical Society.
Records of Colonel St. George Croghan, R. A. Bailey, and others. CSR State of Virginia.
Letters of W. H. Dunham, 36th O.V.I., 1861-62. Manuscript Collections of the U.S. Army Military History Institute, Carlisle Barracks, Penn.
Papers of General Hugh B. Ewing, 30th O.V.I., Manuscript Collections of the Ohio Historical Society, Columbus, Ohio.
Papers of Gen. John B. Floyd and the Army of the Kanawha National Archives.

Letters from the 45th Va. Infantry, Manuscript Collections, WVU, Morgantown, WV.

Papers of Albert George, 30th O.V.I., Manuscript Collections, Ohio Historical Society, Columbus, Ohio.

Papers of Laban Gwinn and the Gwinn Family, 1861-1865. Originals in the possession of Mr. A. N. Gwinn, Grand Rapids, Michigan.

Diary of Samuel Harrison, 44th O.V.I., Manuscript Collections of the Ohio Historical Society, Columbus, Ohio.

Papers of J. W. Hodnett in the manuscript collections of Duke University, Durham, North Carolina. (13th Georgia Infantry)

Diary of C. L. Howard, 13th Georgia Infantry, in the manuscript collections of the Georgia State Archives, Atlanta.

Letters of Nancy Hunt, 1862-1865 in the manuscript collections of the WVU, Morgantown, WV.

Papers & Diary of James Ireland, 12th O.V.I., Ohio Historical Society, Columbus.

Papers of Frank Jones, 13th O.V.I., Cincinnati Historical Society, Cincinnati.

Letters received by the Federal Army Headquarters in the Kanawha Valley from various Union Posts, 1863-1864. Rutherford B. Hayes Library, Freemont, Ohio. Microfilm roll 272.

1861 Papers of Gen. Robert E. Lee, Manuscript Collections of the Va. Historical Society, Richmond, Va.

Papers of William Ludwig, West Virginia University Library, Morgantown, WV.

Diary of Arnold Moss Mason, 16th Tennessee Infantry, September 6, to December 28, 1861. Tennessee State Archives, Nashville.

Diary of Samuel J. Mullins, 42nd Va. Infantry. Unpublished papers in the possession of Mr. R. P. Gravely, Martinsville, Va.

Papers of W. S. Newton, 91st O.V.I., Ohio Historical Society, Columbus.

Papers of the 9th O.V.I., Cincinnati Historical Society, Cincinnati, Ohio.

Papers of George S. Patton, Virginia Military Institute Archives.

Pollock papers, 14th North Carolina Infantry, Southern Historical Soc. Collection, University of North Carolina.

Returns from U.S. Military Posts, 1800-1916, Microcopy no. 617—National Archives.

Papers of C. S. Powell, 14th North Carolina Infantry, S.H.S.C. University of N.C.

Diary of A. B. Roler, Wise Legion, July-September 1861. Virginia Historical Soc. Richmond, Va.

"Diary of a Border Ranger," James D. Sedinger, Co. E, 8th Va. Cavalry, unpublished personal narrative, W.Va. Archives.

Papers of Isaac Noyes Smith, 22nd Va. Infantry, Virginia Historical Soc. Richmond.

Diary of Andrew Stairwalt, 23rd O.V.I., unpublished personal narrative in the manuscript collections of the Rutherford B. Hayes Center, Freemont, Ohio.

Papers of Thomas Taylor, 47th O.V.I., Ohio Historical Society, Columbus, Ohio.

Papers of the Temple brothers, Andrew, George, and Oliver, 1861-1864. Letters written from various W.Va. camps of the 34th, & 89th O.V.I. From originals owned by Ron Meininger, Antebellum Covers, Gaithersburg, Md.

Letters of Gordon Thompson, 60th Va. Infantry. From originals owned by Roger Thompson, 130 W. 11th. Ave. Huntington, WV.

Research papers of the Trans Allegheny Historical Association, Box 65 Beaver, W. Va. Mr. Jody Mays.

Papers of George B. Turner, 91st O.V.I., Ohio Historical Society, Columbus, Oh.

Papers of R.D. VanDuersen, 12th O.V.I., Ohio Historical Society, Columbus, Oh.

Papers of Captain John V. Young, 13th Va. Infantry (U.S.) in the Roy Bird Cook Collection, WVU, Morgantown, WV.

Index

In this index names of the most prominent personalities are not indexed in there entirety, this also applies to frequently used place names.

Abbott, Joel H.—160, 192
Alderson, George—61
Anderson, Frank—26, 70
Anderson, Samuel R.—80
Ansted—106
Bahlmann, W.H.—16
Bailey, E.B.—13
Bailey, Robert A.—17, 153
Barbee, Andrew R.—18
Beckley, Alfred—13, 22, 51, 54, 57-58
Beckley, Henry—18
Beckley, WV.—22, 60, 92
Benham, Henry—85, 90, 108, 112, 115, 120, 123
Bierce, Ambrose—32
Biernes Sharpshooters—24
Big Spring—80, 94
Blakes Farm—116-117
Blue Sulphur Springs—94
Boone County—32
Boone County Rangers—18
Border Guards—18
Border Rangers—18
Braxton County—32
Brock, John P.—21, 24, 46
Broun, Thomas—48
Brown, John—18
Buckholtz Artillery—20
Buffalo Academy—12
Buffalo Guards—18
Bulltown—27
Bungers Mill—31-32, 37

Cabell County—32
Camp Defiance—60, 70, 73, 77, 80, 83, 94, 187
Camp Dickerson—97, 109, 117
Camp Ewing—148
Camp Gauley—45, 51, 55, 59
Camp Huddleston—103
Camp Laurel—58
Camp Lookout—71, 84, 87-88
Camp Reynolds—174, 178, 180
Camp Tompkins Fayette—105
Camp Tompkins Kanawha—13, 15
Camp Two Mile—22
Camp Union—140
Camp Vinton—174, 177, 186
Camp White—181
Cannelton—58, 102, 106
Canty, H.—103
Carnifex Ferry—11, 45-46, 58-59
Caskie Rangers—24, 93
Cassidys Mill—93
Chapman, General—38, 54-55, 86
Chapmanville Daredevils—18
Charleston—16, 20, 25, 28, 55, 148, 162, 170, 181
Charleston Sharpshooters—18
Cheat Mountain—68, 82, 99
Cincinnati—16
Clarksburg—34, 40, 51, 97, 167
Clifton—28
Cliff Top—41
Coal River—13, 45
Coal River Rifle Co.—18

221

Coalsmouth—13, 15
Comstock, Jim—11
Cotton Hill—56, 92, 97, 102, 108, 115, 122, 155, 169, 187
Cox, Jacob D.—23, 29, 38, 40, 45, 51, 59, 74, 87, 107, 133, 147, 170
Croghan, St. George—21, 25, 27, 42-43, 120, 123
Crook, George—171, 180, 190
Cross Lanes—11, 51, 59
Davis, Jefferson—60, 67, 76, 134, 187
Davis, J. Lucious—38, 99, 123
DeVilliers, Colonel—116-117
Dixie Rifles—13, 17, 20, 30
Dogwood Gap—42, 46, 51, 55
Donnelly, Shirley—41
Douglas, Marceleus—103
Dublin, Va.—97, 122, 126, 190
Dunn, Isaac B.—97
Echols, John—169-170
Ector, W.—103
8th Va. Cavalry—192
89th Ohio Infantry—180
11th Ohio Infantry—38, 43, 46, 55, 61, 76, 88, 105, 108, 117
Elk River Tigers—18
Elkwater—71
Ewing, Hugh B.—78
Fayette Rangers—17
Fayette Rifles—16-17
Fayetteville—11, 16, 20, 32, 51, 54, 90, 96, 119, 133, 152, 165, 180, 187
Fife, William—18
5th Va. Infantry U.S.—123, 181
50th Va. Infantry—68
51st Va. Infantry—68, 103, 117, 152-153
1st Illinois Dragoons—133
1st Ky. Infantry—23, 27, 32, 110, 116, 147
1st Tennessee Infantry—80
Flat Top Mtn.—11, 191
Floyd, John B.—31, 37, 42, 48, 55, 65, 73, 82, 90, 99, 105, 110, 115, 123, 126, 167
Floyds Legion—82
Fort Toland—173
14th N.C. Infantry—68, 71, 93
4th Battalion Louisiana Inf.—103
4th Va. Infantry U.S.—123
48th Va. Infantry—80
45th Va. Infantry—43, 68, 86, 103, 109, 152
44th Ohio Infantry—123, 142, 148, 155, 159
42nd Va. Infantry—80, 94
47th Ohio Infantry—119, 123, 133, 155, 165
Frizell, Joseph—38, 46
Gallipolis Ohio—23
Garnet, Robert S.—13
Gauley Bridge—11, 18, 25, 29, 40, 51, 59, 74, 87, 101, 107, 125, 165, 171, 182
Gauley Mount—13, 33-34, 105, 124, 133
Gauley Mountain—20, 88, 102
Georgia Infantry—68, 71, 76, 92, 103, 110
Giles Fayette & Kanawha Tpk.—11
Gillispie, H.L.—11
Glen, Jéan—119, 140, 143
Grant, Thomas—15
Grant, U.S.—190
Greely, Horace—41
Greenbrier County—32, 38, 60, 67, 124, 170

Greenbrier Riflemen—21
Guyandotte River—27
Gwinn, Laban—142-143, 145
Gwinn, Mary—142-143, 145
Hale, John P.—18
Hamilton, A.W.—109
Hamilton, Matilda—43
Hansford, C.M.—15
Harpers Ferry—18
Hawks Nest—30, 42-43, 46, 55, 59, 87
Hayes, Rutherford B.—54, 106, 174, 180
Henningson, Charles—21, 70
Heth, Henry—38, 60, 68, 86, 90, 103, 109, 118, 135
Hico, WV.—42, 45, 59
Hill, Fanny—120
Honey Creek—55
Hot Springs, Va.—86
Hughes Creek—18
Humphreys, M.W.—183-184
Hunt, Nancy—191-192
Hunt, Ralph—57
Imboden, Frank—68, 70
Imboden, John D.—183
Jackson, Andrew—40
Jackson Avengers—24
Jacksons Invincibles—21
Jackson River—77, 93
Jackson, Stonewall—172
James River & Kanawha Tpk.—11, 40, 70, 173, 191
Jenkins, Albert G.—18, 22, 46, 148, 167, 170
Jones, B.H.—13, 17, 20, 30, 152
Jones, Levi—119
Joyce, John—54
Kanawha Artillery—18
Kanawha Falls—18, 32, 51, 56, 78, 105, 134, 159, 167
Kanawha Rangers—13
Kanawha Riflemen—17
Kanawha Salines—58, 170
Kanawha Valley Star—13-14
Kesslers Cross Lanes—11, 45, 51, 59
Kirbys Artillery—24
Lane, Philander P.—105
Lee, Mary C.—86
Lee, Robert E.—12, 27, 37, 46, 58, 73, 82, 91, 96, 99, 108, 167, 183, 192
Lewis, Charles—13
Lewis, Samuel—13
Lewis, Thomas—18
Lewisburg—18, 20, 31, 65, 69, 86, 94, 97, 123, 167, 190
Lightburn, A.J.—147-148, 155, 171
Locust Lanes—40
Logan Court House—94
Logan Riflemen—18
Logan Wildcats—18
Loring, William W.—31, 80, 86, 95, 99, 148, 156, 162, 165
Louborough, Nathan—56-58
Louisiana Rangers—68, 70, 103
Loup Creek—45, 108, 112, 115, 170, 186
Lovers Leap—109
Manassas, Va.—34, 37, 88, 94
Mathews, Mason—67
Matthews, Stanley—54
Maurry, Matthew F.—41
McCausland, John—12, 23, 103, 162, 183, 185, 187

McClellan, George B.—15, 23, 34, 51, 124
McCoys Mill—119, 140, 143
McElroy, J.N.—28
McKinley, William—174
McMullins Artillery—88, 133, 183
Meadow Bluff—38, 60, 65, 74, 78, 93, 96, 124, 191
Millers Ferry—90, 96
Mississippi Infantry—77, 103, 117
Montgomerys Ferry—99, 105, 116, 159, 169, 180
Mountain Cove—87-88, 173
Mountain Cove Guard—17
New River—20, 33, 51, 59, 97, 105
New River Narrows—11, 181
Newton, W.S.—184, 186-187
Nicholas Blues—18
Nicholas County—22, 32, 51, 61, 183
19th Brigade Va. Militia—54
9th Ohio Infantry—109
91st Ohio Infantry—184, 186-187, 190
92nd Ohio Infantry—172, 177
North Carolina—21, 25
North Carolina Infantry—68, 71, 93
Ohio Infantry—23, 38, 42, 51, 56, 61, 76, 105, 109, 123, 133, 140, 147, 155, 174, 184
Ohio Valley—32
Old Stone House—40-41
184th Va. Militia—22, 51, 58
187th Va. Militia—22
142nd Va. Militia—12, 22, 51, 58, 63
190th Va. Militia—22
129th Va. Militia—22
Painter, Steward D.—43, 45
Patton, George S.—17, 22, 153
Pfaus Cavalry—88
Phillips Legion—90, 96, 103
Piggots Mill—43, 46, 191
Pig River Invincibles—21
Pocanhontas County—37, 94, 170
Pocatalico—18
Point Pleasant—27, 32
Prisoners—136-139
Putnam County—27
Putnam County Riflemen—18
Rainelle, WV.—86
Raleigh County—11, 124, 147
Randolph County—68, 74
Ravens Eye—40
Ravenswood—27
Richmond—18, 25, 40, 61, 76, 99, 123, 190
Richmond Blues—21, 24, 68
Rockingham Cavalry—21
Roler, Addison B.—31
Rosecrans, William S.—34, 40, 45, 51, 58, 69, 82, 93, 97, 107, 112, 123, 134, 191
Russell, Dan R.—103
Sandy Rangers—18
Scammon, E.P.—171, 180
Scary Creek—27, 37
Schenck, Robert C.—77, 82, 88, 90, 121
Schutte, John—42
Scrabble Creek—20, 54, 107
2nd Ky. Infantry—23, 27, 51, 54, 57, 87, 102, 105, 107, 117
2nd Va. Cavalry U.S.—123, 148, 155, 180
7th Ohio Infantry—42, 45, 51, 119, 121
7th Tennessee Infantry—80

Sewell Mountain—34, 38, 60, 70, 77, 85, 90, 97, 108, 125, 187, 191
Shackelford, F.G.—153, 159
Shelton, Winston—18
Siber, Edward—147-148, 153, 156
16th Tennessee Infantry—80, 82, 91
60th Virginia Infantry—78, 183-184
Slavery—11, 13, 16
Smith, Isaac N.—71, 83-84, 101, 115, 121
Snake Hunters—43
Spalding, James W.—78
Spy Rock—61, 70, 87
Staunton & Va. Central R.R.—96
Summersville—11, 32, 34, 58, 148, 171, 183, 190
Sunday Road—42-43, 59
Sutton—11, 27, 55, 59
Swann, John S.—18
Swann, Thomas—18
Taylor, Walter H.—71, 74, 79, 85, 99
Teays, S.T.—15
Teays, V.H.—15, 27, 162
Tennessee Infantry—80, 82, 91
13th Georgia Infantry—68, 71, 76, 92, 103, 110
13th Ohio Infantry—119, 121
13th Virginia Infantry U.S.—181, 187
30th Ohio Infantry—78, 133, 140
34th Ohio Infantry—140, 147, 153, 155
37th Ohio Infantry—143, 147, 153
36th Ohio Infantry—121, 123-124
36th Virginia Infantry—68, 103, 109, 121, 153, 159, 183
Thompson, Gordon—184
Thurmond, W.D.—133, 142-144
Tompkins, C.Q.—13, 15, 18, 20, 23, 84, 90, 103, 115
Tompkins, Ellen W.—88, 123
Tompkins Farm—88, 90, 148, 155, 182, 187
Turkey Creek—55
Turner, Charlie—15
Turner, George—172, 177
Turner, Theodore—15
12th Ohio Infantry—23, 184
20th Mississippi Infantry—77, 103, 117
28th Ohio Infantry—109, 123, 133-134
21st Ohio Infantry—23
Twenty Mile Creek—18, 45
22nd Virginia Infantry—68, 71, 77, 101, 103, 109, 115, 121, 123, 152
27th Brigade Va. Militia—56
26th Ohio Infantry—56, 61, 88
23rd Ohio Infantry—54, 133, 174
Tyler, Erastus B.—45, 51
Tyler Mountain—28
Tyree, Francis—13
Tyree, Frank—40, 85, 191
Tyree, Margaret—40, 85
Tyree, Richard—40
Tyree Tavern—40
Tyree, William—17
Valley Mountain—37
Valley Rangers—21, 24
Virginia Infantry—43, 68, 71, 77, 80, 86, 94, 101, 103, 109, 121, 152, 184
Va. Military Institute—12
Virginia Militia—12, 22, 51, 54, 56, 58, 63
Va. Partisan Rangers—106
Virginia & Tennessee R.R.—11, 31, 140, 181, 190

223

Waddell, George C.—103
Warner, Luther—169
Warwick, Bradfute—20
Washington, John A.—71
Wayne Co. Cavalry—18
Weston—11, 27
Wharton, G.C.—103
Wheeling—59
White Sulphur Rifles—24
White Sulphur Springs—37, 94, 99
Wilderness Road—67, 69, 124
Williams, John S.—152, 162
Wise Brigade—48
Wise, Henry A.—18, 22, 25, 37, 40, 46, 55, 59, 65, 74, 133, 186
Wise Legion—21, 23, 31, 46, 68, 82, 93, 99, 123
Wise, Obidiah J.—21, 68
Womack, J.J.—80
Wright, John—22
Wythville, Va.—43, 124
Young, John V.—187

About the Author

Tim McKinney is a native of Fayette County West Virginia and is currently vice president of the Fayette County Historical Society. Stories about his Civil War research and relic hunting have appeared in numerous newspaper and magazine articles and he has appeared on television and radio. Mr. McKinney was guest speaker at the 1986 West Virginia Division Meeting of the United Daughters of the Confederacy, held at Oak Hill, W. Va. He is a member of Camp Garnett, Sons of Confederate Veterans, Camp #1470, Huntington, W. Va., and is a member of the Fraternal Order of Police, Memorial Lodge #118, Fayetteville, W. Va.

Made in the USA
Middletown, DE
23 September 2024